PERGAMON MATERIALS SERIES
VOLUME 9

The Local Chemical Analysis
of Materials

PERGAMON MATERIALS SERIES

Series Editor: Robert W. Cahn FRS
Department of Materials Science and Metallurgy, University of Cambridge, Cambridge, UK

PERGAMON MATERIALS SERIES

The Local Chemical Analysis of Materials

John W. Martin

Department of Materials,
University of Oxford, UK

2003

ELSEVIER

Amsterdam – Boston – Heidelberg – London – New York – Oxford
Paris – San Diego – San Francisco – Singapore – Sydney – Tokyo

Property of Library
Cape Fear Community College
Wilmington, NC

ELSEVIER Ltd
The Boulevard, Langford Lane
Kidlington, Oxford OX5 1GB, UK

First edition 2003
Library of Congress Cataloging in Publication Data
A catalog record from the Library of Congress has been applied for.

British Library Cataloguing in Publication Data
A catalogue record from the British Library has been applied for.

ISBN 0-08-043936-5

∞ The paper used in this publication meets the requirements of ANSI/NISO Z39.48-1992 (Permanence of Paper).
Printed in The Netherlands.

Series Preface

My editorial objective in this Series is to present to the scientific public a collection of texts that satisfies one of two criteria: the systematic presentation of a specialised but important topic within materials science or engineering that has not previously (or recently) been the subject of full-length treatment and is in rapid development: or the systematic account of broad theme in materials science or engineering. The books are not, in general, designed as undergraduate texts, but rather are intended for use at graduate level and by established research workers. However, teaching methods are in such rapid evolution that some of the books may well find use at an earlier stage in university education.

I have long editorial experience both in covering the whole of a huge field – physical metallurgy or materials science and technology – and in arranging for specialised subsidiary topics to be presented in monographs. My intention is to apply the lessons learned in 40 years of editing to the objectives stated above. Authors (and in some instances editors) have been invited for their up-to-date expertise and also for their ability to see their subject in a wider perspective.

I am grateful to Elsevier Ltd., who own the Pergamon imprint, and equally to my authors and editors, for their confidence, and to Mr David Sleeman, Publishing Editor, Elsevier Ltd for his efforts on behalf of the Series.

Here with, I am pleased to present to the public the ninth title in this Series, on a topic of great current concern.

ROBERT W. CAHN, FRS
(*Cambridge University, UK*)

Preface

Any attempt to give an up-to-date account of physical methods of chemical analysis of materials must suffer from the problem of aiming at a moving target. In the chapters which follow I have attempted to illustrate the selected techniques with examples taken from the recent literature of the subject. However I am aware that there is constant instrument development and improvement, so that what follows is at best only a description of analytical equipment that is commercially available at the present time.

I have been fortunate in having been able to impose upon a number of expert friends and acquaintances to read certain relevant chapters of the book. They have given invaluable help in enabling me to correct blatant inaccuracies and to avoid the inclusion of descriptions of wholly obsolete practice. I do, however, accept full responsibility for any remaining errors and omissions in the text which follows.

In particular I wish to thank Professors Alfred Cerezo and John Titchmarsh, Drs Geoffrey Grime and Peter Northover, as well as Dr Wolfgang Bohne and Professor D.B. Williams for their expert advice. I must also express my gratitude to Professor Robert Cahn, who suggested the subject of this book to me a few years ago, for his continued encouragement and support during its preparation. Finally, I thank Professor George Smith of the Department of Materials, Oxford University, for the generous provision of office space in his Department during this period.

John W. Martin

Oxford
January 2003

Acronyms

AEM	analytical electron microxscopy
AES	Auger electron spectroscopy
AP	atom probe
APFIM	atom probe field ion microscope
CAE	constant analysis energy
CHA	concentric hemispherical analyser
CMA	cylindrical mirror analyser
CRR	constant retard ratio
CRT	cathode ray tube
DRIFTS	Diffuse reflectance infrared Fourier transform spectroscopy
DTGS	deuterium triglycine sulphate
ECAP	energy compensated atom probe
EDS	energy dispersive analysis
ELNES	energy-loss near edge structure
EPMA	electron probe X-ray microanalysis
ERDA	elastic recoil detection analysis (see also FreS)
ESCA	electron spectroscopy for chemical analysis
EXELFS	extended energy-loss fine structure
FEG	field emission gun
FIB	focused ion beam
FIM	field ion microscope
FreS	forward recoil spectrometry (see also ERDA)
FTIR	Fourier transform IR spectrometer
FWHM	full width at half maximum
FWTM	full width at tenth maximum
HFS	hydrogen forward scattering
HIBS	heavy ion backscattering spectrometry
HREELS	high resolution electron energy loss spectroscopy
HT	high tension
IBA	ion beam analysis
IR	infra red
IRAS	infrared reflection absorption spectroscopy (see RAIRS)
LAMMA	laser-microprobe mass analysis
LAMPAS	laser mass analyser for particles in the airborne state
LIMA	laser ionisation mass analysis
LIMS	laser ionisation mass spectrometry
MBE	molecular beam epitaxy

MCA	multichannel analyser
MDL	minimum detection limit
MDM	minimum detectable mass
MIR	Multiple Internal Reflection Spectroscopy
MMF	minimum mass fraction
NRA	nuclear reaction analysis
PEELS	parallel electron energy-loss spectrometry
PIGE	proton-induced gamma-ray emission (or PIGME)
PIXE	particle-induced Xray emission
PLAP	pulsed laser atom probe
PoSAP	position sensitive atom probe
PM	polarization modulation
PMT	photomultiplier tube
PSD	position sensitive detector
PTFCE	polytrifluorochloroethylene
RAIRS	reflection absorption infra red spectroscopy (see IRAS)
RBS	Rutherford Back-Scattering Spectrometry
RF	radio frequency
RSFs	relative sensitivity factors
SAM	Scanning Auger Microprobe
SEM	secondary electron microscopy
SIMS	secondary ion mass spectrometry
SMALDI	scanning microprobe matrix-assisted laser desorption ionisation
SNMS	secondary neutral mass spectrometry
SSIMS	static secondary ion mass spectrometry
STEM	scanning transmission electron microscopy
TEM	transmission electron microcopy
TIR	transmission infrared
TOF	time-of-flight
UHV	ultrahigh vacuum
VELS	vibrational energy loss spectroscopy (alternative name for HREELS)
XEDS	X-ray energy dispersive spectrometry
XPS	X-ray photoelectron spectroscopy
r	back-scattered electrons term
λ	Auger electron attentuation length
3DAP	three-dimensional atom probe

Introduction

In order to understand the relationship between the properties of a material and its structure, which is the *raison d'être* of the materials scientist, three important experimental areas of investigation may be necessary. Firstly, of course, the *physical or mechanical properties* in question must be measured with maximum precision, then the *structure* of the material must be characterised (this itself may refer to the *atomic arrangement or crystal structure*, the *microstructure*, which refers to the size and arrangement of the crystals, or the *molecular structure*). Finally, the *chemical composition* of the material may need to be known.

THE SCALE OF CHEMICAL ANALYSIS

Bulk analytical data are usually made available by the particular material manufacturer, such as the specification of a particular metal, alloy, ceramic or polymer. This often includes an indication of the maximum levels of impurities that may be present. There are numerous conventional analytical techniques which may be employed to provide these data, and they usually involve the analysis of a relatively large volume of the material in question in order that local heterogenieties do not affect the result.

The properties of materials are often critically dependent upon how the elements defined by the bulk analysis are distributed within the structure of the material, and this factor may require that a local analysis be conducted. There are three types of locality we shall consider:

(a) *Surface analysis* is concerned with the chemical composition and local arrangement of the atoms that make up the top few layers of the surface of a sample. There are numerous examples where a detailed knowledge of surface composition is of practical importance.

All materials will, to some degree, be subject to corrosion and oxidation by their environment, and the critical early stages of attack can often be understood through the use of surface analytical techniques. A similar approach is required to gain an understanding of the fundamental and applied aspects of surface catalysis, which is of great importance in the petrochemical industry.The microelectronics industry has also contributed to the development of modern surface analytical techniques, where there is a necessity to analyse dopant concentration profiles while retaining lateral resolution on the device of better than one micron.

(b) *Analysis at Interfaces* is of particular importance in those cases where the presence of the interface itself causes a change in the composition (and hence the properties) of the material in question. An obvious example is the enhanced concentration of impurity elements observed at grain boundaries in steels and other alloys, which may lead to brittle intergranular failures in the component. Another example is the formation of impurity 'atmospheres' along dislocations in crystals. There are situations where such segregation may be exploited beneficially, so its detection and measurement is of equal importance in this context.

(c) *Analysis of Microstructural Phases*, which may be beneath the surface of the specimen, is the remaining aspect of local analysis to be considered. The size of the individual phases constituting the microstructure of solids can range from the order of microns to nanometres, and our main approach to this topic will be concerned with the use of the various forms of analytical electron microscopy (AEM). The analysis of isolated individual solid particles which may be of submicron size (for example, aerosol particles) can be carried out by techniques other than AEM, and this application will also be considered in the relevant context.

THE APPROACH ADOPTED IN THE TEXT

In the text which follows, our approach has been to review the techniques for local analysis in terms of the *nature of the probe* employed, as follows:

X-ray probes for surface analysis are used in X-ray photoelectron spectroscopy (XPS), and examples are given of a wide range of applications of this technique in materials science.

Infrared and ultraviolet probes for surface analysis are then considered.The applications of IR spectroscopy and Raman microscopy are discussed, and a brief account is also given of laser-microprobe mass spectrometry (LAMMA).

Ion beam probes are used in a wide range of techniques, including Secondary Ion Mass Spectroscopy (SIMS), Rutherford backscattering spectroscopy (RBS) and proton-induced X-ray emission (PIXE). The applications of these and number of other uses of ion beam probes are discussed.

High resolution analytical electron microscopy (HRAEM) is not confined to surface analysis, and applications of this as well Auger (AES) and electron energy loss (EELS) spectroscopies are described.

   The opening chapter, however, gives an account of the ultimate technique in local analysis, namely that of the *atom probe* field ion microscope. Here no probe is employed as in the above list, but individual atoms of the specimen are removed and identified by mass spectroscopy.

Contents

Chapter 1
Atom Probe Microanalysis

Chapter 1
Atom Probe Microanalysis

The *atom probe (AP)* consists of a modified *field ion microscope (FIM)*, and we will first review the basic phenomena involved in the operation of the FIM.

1.1. THE FIELD ION MICROSCOPE

Müller (1951, 1956) developed this instrument, which for the first time enabled extensive details of the atomic structure of a solid surface to be seen directly. Figure 1.1 illustrates schematically the basic construction of a FIM. The specimen is prepared in the form of a fine wire or needle, which has been chemically or electrochemically polished to a sharp point with an end radius typically 50–100 nm. It is mounted along the axis of a vacuum chamber, about 50 mm from a phosphor screen (perhaps 75 mm in diameter). The specimen is mounted on an electrical insulator within a cryostat, and it can be raised to a high positive potential (3–30 kV) by means of the leads attached.

The microscope is filled to a pressure of $\sim 2 \times 10^{-3}$ Pa with an inert gas such as helium or neon, and gas atoms are polarized by the inhomogeneous electric field and drawn towards the apex of the tip. Here, a complex series of processes occur. Firstly, after they become thermally accommodated to the sepcimen temperature, field adsorption of gas atoms occurs on prominent surface atoms. If the field is high enough, atoms beyond a certain critical distance from the surface may become field ionized, and the resulting positively charged gas ions are repelled from the specimen surface towards the fluorescent screen where they build up a highly magnified, projected image of the surface at which they were formed. The narrow beams of ions formed above prominent surface atoms give rise to individual image spots on the phosphor screen.

The efficiency with which the ions convert their energy to light on striking a phosphor screen is only 0.1–1.0%, so the light produced by the direct image is very weak. A microchannel-plate image converter is therefore placed just in front of the screen (see Figure 1.1), which converts the incident ion beams to more intense secondary electron beams, enabling the FIM images to be seen under subdued lighting.

Figure 1.2 shows such an FIM image of tungsten containing a grain boundary, and it consists of a large number of spots arranged in intersecting sets of concentric rings. The rings correspond to the atomic terraces of prominent

3

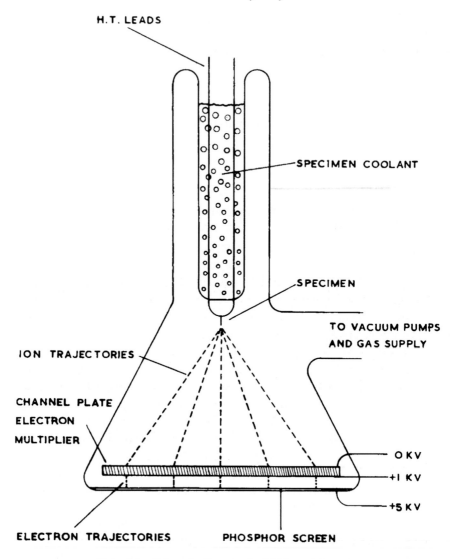

Figure 1.1.   Schematic diagram of a field ion microscope.

crystallographic planes on the specimen surface, and the bright spots correspond to ionization above the edge or corner atoms of these planes. The pattern of the image resembles a stereographic projection of the surface and can be indexed by standard techniques, and the position of the boundary (B) between the two grains which form the specimen is evident from the discontinuity in regularity of the image.

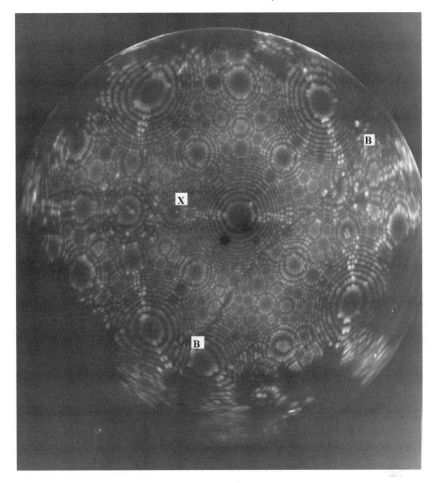

Figure 1.2. FIM image of a tungsten specimen with (110) centre pole A grain boundary runs between B–B, and a single atom is apparent on the plane below the mark X. (Courtesy T.J. Godfrey.)

The magnification of the image (M) obtained for the idealised case where the tip of the specimen and the screen consist of two concentric spheres of separation a is given by

$$M = a/r_0$$

Where r_0 is the apex radius of the specimen. In practice the ion trajectories are curved, due to the presence of the specimen shank, and the magnification is reduced by an image compression factor, β, and

$$M = a/\beta r_0$$

where β is ~ 1.5. For typical values of a (50 mm) and r_0 (35 nm), M is ~ 1 million.

With this imaging system it is possible to study virtually all metals and alloys, many semiconductors and some ceramic materials. The image contrast from alloys and two-phase materials is difficult to predict quantitatively, as the effects of variations in chemistry on local field ion emission characteristics are not fully understood. However, in general, more refractory phases image more brightly in the FIM. Information regarding the structure of solid solutions, ordered alloys, and precipitates in alloys has been obtained by FIM.

1.1.1 Specimen Preparation

A detailed account of methods of AP specimen preparation has been given by Miller and Smith (1989) to which reference may be made if necessary.

A typical blank is about 10 mm long and 0.25 mm square or circular in cross-section, produced from bulk material by standard forming methods. Starting material in the form of thin ribbons or surface layers would require coating the desired region with a photoresist film, then writing a suitable pattern followed by etching away the unwanted material.

The blank is then normally electropolished or chemically etched into a sharp point. The two-stage electropolishing process illustrated in Figure 1.3 is commonly employed.

In the first stage a thin layer of electrolyte is floated on a dense inert liquid such as carbon tetrachloride. The specimen is moved slowly up and down during electropolishing to minimize preferential attack at the air–electrolyte interface and to form a necked region on the specimen blank. The second stage continues to remove material until the weight of the lower half is too great to be supported by the neck, thus producing two FIM specimens.

In the case of more difficult specimens, focussed ion beam milling may be employed in their preparation, and discussed in Section 1.7.

1.2. THE CONVENTIONAL (ONE-DIMENSIONAL) ATOM PROBE

An FIM may be modified so that the imaged atom chosen for analysis can be positioned over a small aperture in the phosphor-coated screen. If the electric field is raised to a sufficiently high value, material may be removed from the surface by field evaporation. The specimen is subjected to a high-voltage pulse, which causes a number of atoms on the specimen surface to field evaporate as positive ions. Only the atom that was imaged over the aperture (or 'probe hole') passes into a time-of-flight mass spectrometer, all the other atoms being blocked off by the screen. The applied

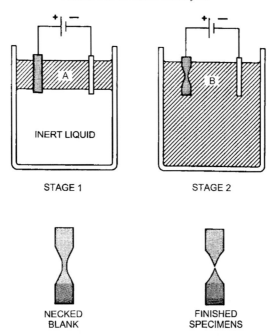

Figure 1.3. Illustration of the two-stage electropolishing process for the production of FIM specimens. In Stage 1 a 5–7 mm thick layer of electrolyte is floated on a denser inert liquid, producing a necked region in the centre of the blank. In Stage 2, polishing is continued until the weight of the lower half of the specimen blank is too heavy to be supported by the thin neck. (After Miller *et al.* 1996.)

pulse starts computer-controlled digital timing systems; the timing is stopped when the ion is recorded as striking the single atom sensitive detector. By coupling the FIM with a time-of-flight mass spectrometer, it is possible to determine the mass-to-charge ratio of individual ions.

In this way, Müller *et al.* (1968) were the first to incorporate a small probe hole into the FIM image screen that formed the entrance to a time-of-flight mass spectrometer that had single atom sensitivity. The specimen was mounted on a stage such that a small area of the image could be selected for analysis by aligning its field ion image with the aperture, and so the identification of selected individual atoms became possible for the first time. The AP may thus determine the chemistry of a nanometre-scale region by simply counting the number of atoms of each type in a given volume.

The mass-to-charge ratio, m/n, of those ions that pass through the probe aperture and are analysed in the mass spectrometer is calculated from the equivalence between the potential energy of the atom on the specimen surface at voltage V_0, and the kinetic energy that the atom acquires during acceleration to the grounded counter electrode. The ions are considered to acquire their final velocities instantaneously, hence

$$neV_0 = (1/2)m(d^2/t^2)$$

i.e. $$m/n = 2eV_0(t^2/d^2)$$

where d is the distance travelled by the ion from the specimen to the detector, e is the elementary charge and t is the flight time of the ion over that distance.

Since only one or two charge states are observed for any element, the elemental identity of each ion is determined from its mass-to-charge ratio by consulting a table of known isotope abundances.

1.3. ENERGY COMPENSATED ATOM PROBE (ECAP)

The ECAP incorporates an electrostatic lens in the time-of-flight spectrometer in order to improve the mass resolution by compensating for small spreads in the energies of the ions evaporated from the specimen under the pulsed electric field. A lens design by Poschenrieder or a reflectron type of electrostatic lens is used for this purpose, and is standard equipment for metallurgical or materials applications of APFIM. These typically improve the mass resolution at full width half maximum (FWHM) from $m/\Delta m \sim 250$ to better than 2000.

A schematic diagram of the conventional atom probe due to Hono (1999) is shown in Figure 1.4.

The data chain of the collected atoms can be converted to a one-dimensional composition-depth profile. The depth profile shows an average concentration of solute within the aperture, and there is always a possibility that the chemical information from the selected area is a convolution of more than one phase, as indicated diagrammatically in Figure 1.5, which represents the analysis of a FIM specimen containing second phase particles and also an interface across which there is a change of composition.

This schematic diagram shows that the true composition of particles can be obtained only when the probe hole covers the particle entirely. When the probe hole covers both the particle and the matrix, the measured concentration is lower than the real one. Again, when an interface is not perpendicular to the cylinder of analysis, the apparent concentration change at the interface appears diffuse, even if the real concentration change is discrete. The standard deviation for concentration, σ, is given by

$$\sigma = \sqrt{\frac{x(1-x)}{n}}$$

where n is the number of atoms and x is the atomic fraction of the solute.

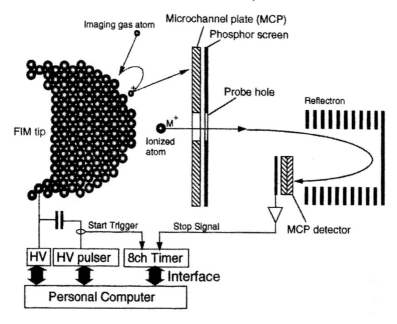

Figure 1.4. Schematic diagram of the basic principles of the conventional (one-dimensional) atom probe, with a reflection energy compensator. (Reproduced by permission of Hono 1999.)

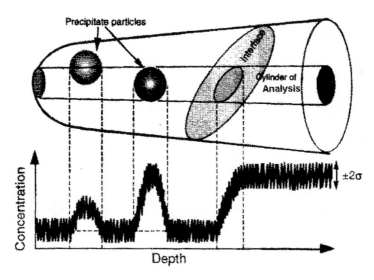

Figure 1.5. Schematic drawing illustrating how the one-dimensional atom probe analyses particles and interfaces in a FIM specimen. (Reproduced by permission of Hono 1999.)

1.4. THE THREE-DIMENSIONAL ATOM PROBE (3DAP)

The 3DAP permits the elemental reconstruction of a small volume of the specimen with near atomic resolution, by determining the x, y and z positions and mass-to-charge ratios of the atoms detected in that volume. Early instruments were built without an energy compensating lens, resulting in a lower mass resolution ($m/\Delta m \sim 50$–250). However, a reflectron lens was later incorporated into a 3DAP instrument by Cerezo *et al.* (1998), giving an improvement of mass resolution to $m/\Delta m \approx 600$.

A position sensitive detector (PSD) is employed, of which there are several types used effectively around the world. One type is essentially a square array of multi-anodes, as shown in Figure 1.6. By measuring the time-of-flight and the coordinates of the ions upon the PSD, it is possible to map out a two-dimensional elemental distribution. The elemental maps are extended to the z-direction by ionizing atoms from the surface of the specimens. The z position is inferred from the position of the ion in the evaporation sequence, so that the atom distribution can be reconstructed in a three-dimensional real space.

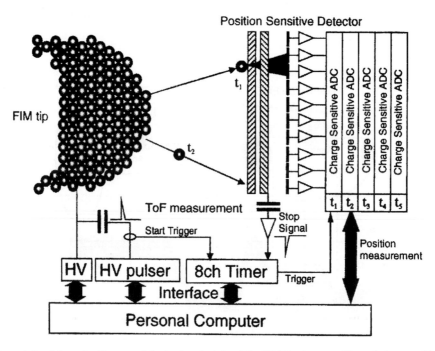

Figure 1.6.   Schematic diagram of the basic principle of the 3DAP using a multi-anode type position-sensitive detector. (Reproduced by permission of Hono 1999.)

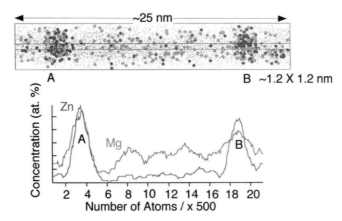

Figure 1.7. 3DAP reconstruction of the solute distribution in Al-1.7Zn-3.4Mg alloy following ageing at 90°C for 20 hr. The pale spheres represent Zn atoms, darker spheres Mg atoms and fine black dots represent Al atoms. (Reproduced by permission of Ringer and Hono 2000.)

Figure 1.7 illustrates such a three-dimensional reconstruction of the solute distribution in an Al-1.7at%Zn-3.4at%Mg alloy following ageing at 90°C for 20 h (Ringer and Hono 2000). These data suggest that the Zn and Mg atoms have formed spherical GP zones of approximately equal Mg–Zn composition.

1.5. ANALYSIS OF INTERFACES BY AP

Because of the small field of view of the APFIM, the chances of finding a grain boundary are low, so TEM of FIM tips is carried out in order to preselect specimens which contain interfaces. The mechanical stress on the specimen arising from the strong electric field in the FIM may result in fracture of embrittled specimens, so that the AP is preferred in situations where segregation has not resulted in significant embrittlement of the material. The technique may be regarded as complementary to AES, in which interfacial embrittlement is a prerequisite for its application.

An extensive study of boron segregation to grain boundaries in 316L stainless steels has been performed by Karlsson and Nordén (1988). The size of the boron atoms (larger than the common interstitial elements in steel, e.g. carbon and nitrogen) and smaller than substitutional elements (e.g. chromium and nickel) implies that it should be energetically favourable for boron to migrate to loosely packed regions like interfaces. It has long been known that the addition of small amounts of boron to austenitic steels improves their hot workability and creep resistance, and it was believed that segregation of the element to the grain boundaries was responsible.

The geometry of APFIM analysis of interfaces is illustrated in Figure 1.8. During analysis the interface moves across the probe hole, and the increase in number of solute atoms per unit area of the interface (Γ) can be calculated.

If N_i is the total number of detected atoms of the segregating element i, N the total number of atoms detected, c_m the matrix concentration of the element, then

$$\Gamma = (N_i - Nc_m)/A\eta$$

where A is the probed area of the interface, given by $\pi r^2/\cos\phi$, where r is the radius of the probe hole and η is the detection efficiency of the ion detector (50–90%).

Karlsson and Nordén (1988) observed nonequilibrium segregation of boron in their steel arising from the transport of vacancy-solute pairs to grain boundaries during cooling. They employed three cooling rates from 0.29 to $530\,\mathrm{C\,s^{-1}}$ for three starting temperatures: 800, 1075 and 1250°C. The total amount of segregation was highest at intermediate cooling rates, that is, when time was sufficient to let vacancy-B pairs diffuse to the grain boundary but not enough to let B diffuse away from the enriched zone. The grain size of the steel was about $200\,\mu\mathrm{m}$, which made specimen preparation very demanding. Figure 1.9 illustrates their atom probe results from a specimen containing 206 ppm boron which had been cooled at $0.29\,\mathrm{C\,s^{-1}}$ from 1250°C.

The binding energy of boron in austenite grain boundaries was estimated at 0.65 eV. The influence of the relative grain orientation on the amount of non-equilibrium segregation seemed to be small for general high and low-angle boundaries, although no segregation was detected at coherent twin boundaries.

Thuvander and Andrén (2000) have reviewed APFIM studies of grain and phase boundaries and demonstrate that the technique has played a vital role in the understanding of interfacial chemistry in many important materials including

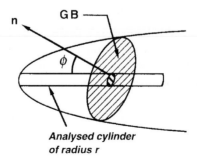

Figure 1.8.   The geometry of APFIM analysis of interfaces. ϕ is the angle between the analysis direction and the normal (n) to the grain boundary.

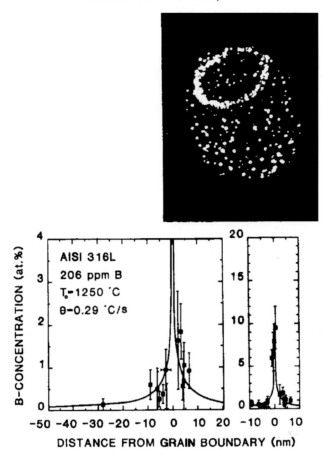

Figure 1.9. FIM micrograph showing boron segregation to a grain boundary in austenitic 316L stainless steel, together with atom probe composition profiles for boron (Karlsson and Nordén 1988).

a range of metals and alloys, intermetallics, cemented carbides as well as internal metal/oxide interfaces. In a few specific cases, overlaps in the mass spectrum makes it difficult to distinguish between two possible segregants, for example, N-Si and O-S. They conclude that APFIM is by far the best method for studying enrichment of segregated atoms at grain boundary dislocations, the segregation and trapping of deuterium, segregation to nanometre-sized precipitates, segregation of light elements (B,C,N,O) to nonbrittle boundaries and segregation of C to boundaries decorated with carbides.

The development of the 3DAP has made it possible to analyse much larger areas of an interface, which improves the statistics as well as the detection limit. Thomson and Miller (2000) provide an impressive example of this technique, as shown in

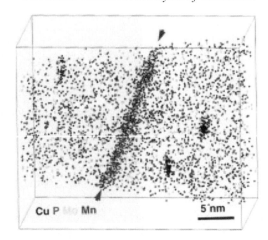

Figure 1.10. Grain boundary segregation in neutron irradiated weld from Russian VVER 440 nuclear reactor: position of individual atom is represented by a dot. Enrichment of P, Mo and Mn atoms are visible at boundary (arrowed), and three 2 nm diameter Cu-enriched precipitates near the boundary are also visible. (Reproduced by permission of Thomson and Miller 2000.)

Figure 1.10. This shows a 3DAP atom map across a grain boundary in a neutron irradiated 15KkMFA Russian pressure vessel steel weld. It reveals significant levels of P, Mn, Mo and C segregation, which correlate with poor mechanical properties exhibited by these materials.

1.6. ATOM PROBE STUDIES OF SEMICONDUCTOR MATERIALS

Voltage-pulsed AP analysis of resistive materials is difficult, since they do not transmit high-voltage pulses effectively. In the case of semiconducting materials there is the further problem of their brittleness, so that the mechanical shock due to the voltage pulse often causes specimen fracture. The pulsed laser atom probe (PLAP) is more effective for these materials.

The PLAP applies a short duration (100 ps–10 ns) laser pulse to the apex of the specimen. The heat generated is sufficient to promote the field evaporation at the standing voltage of the specimen. The specimens need only to be sufficiently conductive to permit field ion imaging. The peak temperature in the PLAP is only ~300 K for a period of a few nanoseconds, which is not sufficiently high for surface diffusion on semiconductor materials, and so the spatial resolution is not downgraded.

Miller *et al.* (1996) give examples of APFIM studies of the surface oxides formed on silicon, and studies of the stoichiometry of thermal oxide layers of this element, as well as a wide range of binary III–IV semiconducting materials.

1.7. STUDIES OF THIN FILMS

For the successful application of 3DAP to the study of interfacial layers and devices, a specimen preparation technique must be employed which allows needle-shaped samples to be fabricated from materials grown directly on planar substrates. Larson *et al.* (2000b) employed a silicon substrate upon which a large number of columns had been etched. Layers were then deposited and then capped with a protective metal film. Individual columns are broken off for mounting on the end of a pin, and focussed ion beam milling was employed to sharpen the end of the resulting post, as shown in Figure 1.11 to make a specimen of the required sharpness.

The metal capping layer protects the layers of interest from damage during the milling stage. Figure 1.12 shows the 3DAP analysis of a NiFe/CoFe/Cu/CoFe multilayer speciemn prepared by this method, including part of the capping layer and the first two repetitions of the multilayer.

In Figure 1.12(b) the individual atomic planes within the film are shown, together with the changes in chemistry across the four interfaces included in the section. It can be seen, for example, that the CoFe/Cu interfaces are sharper than those between NiFe and CoFe layers. However, the interface with CoFe growing on NiFe shows a more diffuse interface than that where NiFe is deposited on CoFe.

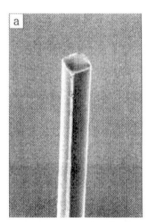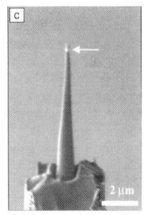

Figure 1.11. (a)–(c) Sequence of secondary-ion images during the preparation of a 3DAP specimen from a multilayer thin film. (a) Oblique view of the silicon post with layers deposited on to the flat end. (b) Similar view of the specimen sharpened by focussed ion-beam (FIB) milling. (c) Side view showing the specimen more clearly: note the bright contrast from the metallic layers at the specimen apex (arrowed). (Courtesy Cerezo *et al.* 2001.)

Figure 1.12. (a) Reconstruction of the 3DAP analysis of a NiFe/CoFe/Cu/CoFe multilayer thin film structure. The Ni, Co and Cu atom positions are shown by green, blue and red dots, respectively. The height of the volume shown is approximately 35 nm. (b) Magnified view from a section of the analysis showing individual (111) planes along the growth direction. (Courtesy Larson *et al.* 2000a.)

1.8. ADSORPTION AND SURFACE REACTIONS

Atom probe techniques have been used to investigate adsorption processes and surface reactions on metals. The FIM specimen is first cleaned by the application of a high-voltage field evaporation pulse, and then exposed to the gas of interest. The progress of adsorption and surface reaction is monitored by the application of a second high-voltage desorption pulse and a controlled time delay.

The use of the PLAP in adsorption studies has been reviewed by Tsong (1990). A related field of investigation is that of catalysis, and any structural and/or morphological changes on the surface of catalyst materials can be followed on the atomic scale by FIM methods. The size and shape of the apex region of an FIM specimen resembles that of a single particle of a supported metal catalyst. The behaviour of gaseous adsorbates on individual crystal planes can be observed as a function of time and temperature, and the power of the technique is illustrated by a number of examples given by Miller *et al.* (1996).

REFERENCES

Cerezo, A., Larson, D.J. & Smith, G.D.W. (2001) *MRS Bulletin* **26**(2) 102.
Cerezo, A., Godfrey, T.J., Sijbrandij, Warren, P.J. & Smith, G.D.W. (1998) *Rev. Sci. Instruments*, **69**, 49.

Hono, K. (1999) *Acta mater.*, **47**, 3127.

Karlsson L. & Nordén, H. (1988) *Acta Metall.*, **36**, 13.

Larson, D.J., Cerezo, A., Martens, R.L., Clifton, P.H., Kelly, T.F., Petford-Long, A.K. & Tabat, N. (2000a) *Appl. Phys. Letters*, **77**, 726

Larson, D.J., Martens, R.L., Kelly, T.F., Miller, M.K. & Tabat, N. (2000b) *J. Appl. Phys.*, **87**, 5989.

Miller, M.K. & Smith, G.D.W. (1989) *Atom Probe Microanalysis: Principles and Applications to Materials Problems*, Materials Research Society, Pittsburgh, PA, USA.

Miller, M.K., Cerezo, A., Hetherington, M.G. & Smith, G.D.W. (1996) *Atom Probe Field Ion Microscopy*, Oxford University Press, Oxford, UK.

Müller, E.W. (1951) *Z. Phys.*, **131**, 136.

Müller, E.W. (1956) *J. Appl. Phys.*, **27**, 474.

Müller, E.W., Panitz, J.A. & McLane, S.B. (1968) *Rev. Sci. Instrum.*, **39**, 83.

Ringer, S.P. & Hono, K. (2000) *Materials Characterization*, **44**, 101.

Thomson, R.C. & Miller, M.K. (2000) *Materials Science and Technology*, **16**, 1199.

Thuvander, M. and Andrén, H.-O. (2000) *Materials Characterization*, **44**, 87.

Tsong, T.T. (1990) *Atom-Probe Field Ion Microscopy*, Cambridge University Press, Cambridge, UK.

Chapter 2

X-Ray Probes for Surface Analysis (XPS or ESCA)

Chapter 2

X-Ray Probes for Surface Analysis (XPS or ESCA)

One important use of X-ray probes is in the study of local order and displacements, but this is not within the scope of the present book. The recent availability in intense synchrotron sources with selectable X-ray energies permits high-precision measurements of chemically specific atomic-pair correlations in solid solution alloys. A recent review of the technique is given by G.E. Ice and C.J. Sparks (Modern Resonant X-ray studies of alloys: local order and displacement) in Annual Reviews of Materials Science 1999, **29**, 25–52.

We shall concern ourselves here with the use of an X-ray probe as a *surface analysis* technique in *X-ray photoelectron spectroscopy (XPS)* also known as *Electron Spectroscopy for Chemical Analysis (ESCA)*. High energy photons constitute the XPS probe, which are less damaging than an electron probe, therefore XPS is the favoured technique for the analysis of the surface chemistry of radiation sensitive materials. The X-ray probe has the disadvantage that, unlike an electron beam, it cannot be focussed to permit high spatial resolution imaging of the surface.

2.1. SAMPLE PREPARATION FOR SURFACE ANALYSIS

A typical sample would have a surface area of the order $10 \times 20\,\text{mm}$. The preparation of the surface for examination is the subject of an ASTM standard (ASTM E1078-97), and the objective is to ensure that the surface to be analysed has not been contaminated or altered prior to analysis. The techniques of surface analysis are sensitive to surface layers only a few atoms thick, so the degree of cleanliness required will be much greater than for other forms of analysis. Nothing must be allowed to touch the surface to be analysed.

Samples may either be those in which the surface of interest has been exposed to the environment before analysis, or the surface to be examined is created in the UHV chamber of the instrument. The latter method is generally preferable, and also argon-ion bombardment is commonly used to clean sample surfaces *in situ* in the spectrometer. In metallurgical studies, the fracture sample is particularly important: the sample is machined to fit the sample holder, and a notch is cut at the desired point for fracture. The fracture stage is isolated from the analytical chamber and is pumped down to UHV. Liquid nitrogen cooling is often provided, as this encourages brittle fracture, and after fracture by impact the specimen is transferred to the analysis chamber and the analysis performed before the newly created surface has become contaminated from the residual gases present in the system.

As mentioned above, argon-ion sputtering may also be employed to clean the surface prior to analysis. A carefully controlled leak valve introduces high-purity argon ($\not< 99.9\%$ purity) at a pressure of 5×10^{-5} torr into a 'gun' operating at about 1 keV. Care must be taken in employing this technique, however, as pointed out by Barr (1994), since many surfaces suffer degradation due to differential sputtering during the sputter etching process. Most of this degradation is selective (depending on the chemistry and physics of the surface), but if the resulting surface can be analysed by XPS, valuable information about the original surface may still be obtained.

As in the case of AES, it is possible to employ sputter etching with XPS to profile a material in depth, although it is a much less straightforward process in the case of XPS. For sequential sputter etching during AES, the electron gun employed for AES need not be turned off during ion etching. However, it is not possible to operate an ion gun and X-ray source in the same chamber, so the process has to be conducted in a side-chamber of the apparatus in two time-consuming stages, making the analysis much more laborious than with AES.

2.2. X-RAY SOURCES

Synchrotron radiation is not available from laboratory sources, and expensive central facilities must be used. The vast majority of XPS spectra are obtained using conventional X-ray tubes as X-ray sources. X-rays are generated by bombarding a metal target (in XPS experiments these are usually of magnesium or aluminium) with electrons of ~ 15 keV energy. This produces a broad band of photons with the characteristic K_α and less intense K_β peaks superimposed (e.g. Figure 2.1). It is preferable to use a monochromatic beam of photons, as this simplifies the X-ray electron spectrum. This is achieved by incorporating quartz crystals to diffract the X-ray beam such that only the K_α X-rays satisfy the Bragg condition and are diffracted on to the sample. The quartz crystal is elastically bent to a radius of curvature such that the diffracted monochromatised beam is focussed upon the specimen surface.

The X-ray sources used in XPS are of longer wavelength and thus of lower energy than that of Mo illustrated in Figure 2.1: the energy of the K_α-line of Mg is 1253.6 eV, and that of Al is 1486.6 eV. The K_α-lines of Ti (2040 eV) are also sometimes used. A twin-anode source is often employed which enables either Mg or Al K_α X-rays to be produced by energising the appropriate filament. The tube and sample have to be separated by a thin film, through which the X-rays pass, to prevent scattered electrons from the tube from penetrating into the sample chamber.

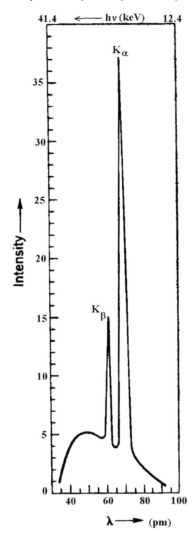

Figure 2.1. The X-ray spectrum of molybdenum, showing K_α and K_β lines superimposed upon the continuous spectrum. The quantum energy is shown in the upper scale; the intensity is in arbitrary units.

2.3. THE PHOTOIONISATION PROCESS

XPS must be carried out in UHV conditions: pressures in the 10^{-10} mbar range are required if contaminants are to be kept below a few per cent during the course of a typical experiment. Photons with energy $h\nu$ produced from the X-ray source are incident upon the sample surface. These are absorbed by the atoms in the surface

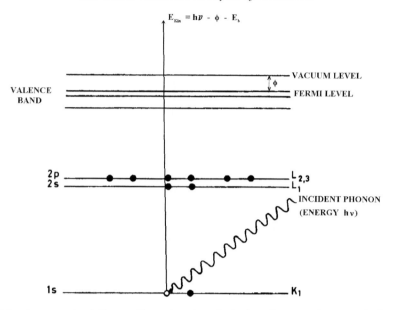

Figure 2.2. An energy level diagram illustrating photo-ionisation of an atom by removal of a K-shell electron. The Fermi level represents zero of the binding energy and the vacuum level represents zero of kinetic energy.

which are ionised by the removal of an inner or outer shell electron. In the example shown in Figure 2.2, the atom is ionised by the removal of a 1s electron from the K electron shell, and the energy of the ejected electron E_{Kin} is given by:

$$E_{Kin} = h\nu - \phi - E_b \qquad (2.1)$$

where ν is the frequency of the incident photon, h is Planck's constant, ϕ is the work function of the surface and E_b is the binding energy of the electron within the atom. Photoelectrons are emitted from all energy levels of the target atom and hence the electron energy spectrum is characteristic of the emitting atom type. The principle of XPS is therefore to measure the kinetic energy of the electrons emitted by an atom absorbing an incident photon with a known energy $h\nu$, so that the binding energy E_b may be calculated by means of equation (2.1). The best energy resolution should be used compatible with the signal-to-noise ratio in the particular spectrum, and this is typically of the order of 1.0 eV. To avoid limitation of the resolution by the line width of the source, it is necessary to use target materials whose characteristic X-ray lines have widths less than 1.0 eV. This requirement is fulfilled by both Mg K_α (line width 0.7 eV) and Al K_α (line width 0.85 eV).

The analysis is applicable to all elements except hydrogen in solids, liquids and gases, although it is normally confined to the study of solids in the form of powder or plates.

The XPS technique is highly surface specific due to the short range of the photoelectrons that are excited from the solid: detection of electrons gives information from a region between 1.5 and 4.0 nm in from the surface. The minimum sample mass that can be analysed is 10^{-8} kg, while 10^{-10} to 10^{-12} kg of an element can be detected. Usually samples of 10–100 mg are taken.

2.4. INSTRUMENTATION

To detect the electrons emitted from the sample surface and to analyse their energy, the electrons are passed through an electron optical component employing electrostatic deflection analysers which spatially disperse electrons across some detection device according to their kinetic energy. Maximum signal intensity is obtained when electrons are collected from the whole of the broad emitting area of the specimen and analysed with a constant resolution generally in the range 0.5–2 eV.

The energy of the photoelectrons leaving the sample are determined using a Concentric Hemispherical Analyser (CHA), and this gives a spectrum with a series of peaks whose energy values are characteristic of each element. A schematic diagram of a CHA is shown in Figure 2.3.

A CHA consists of two metal hemispheres arranged such that their centres of curvature are coincident. Different voltages are placed on each hemisphere, so that an electric field exists between them. Electrons are injected into the gap between the hemispheres, and high-energy electrons will impinge on the outer hemisphere while low-energy electrons will be attracted to the inner hemisphere. Only electrons in a narrow energy region (called the pass energy) succeed in reaching the detector. A series of lenses are placed before the CHA which focus the electrons and slow them down.

There are two operating modes known as Constant Retard Ratio (CRR) or Constant Analysis Energy (CAE). In CRR, the electrons are slowed down by an amount which is a constant ratio of the electron energy to be analysed. For example, if the retard ratio is 10 and 1000 eV electrons are to be detected, then the electrons will be slowed down to 100 eV and the pass energy will be set to 100 eV. In CAE, the pass energy is fixed. If the pass energy is 50 eV, then electrons of 1000 eV will have to be slowed down by 950 eV in order to be detected.

The CRR mode gives constant resolving power and the CAE mode gives constant energy resolution. The main advantage of XPS over AES is its greater sensitivity.

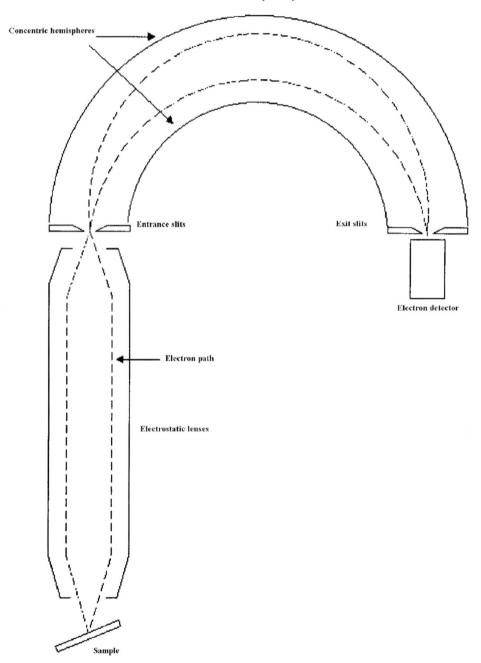

Figure 2.3. Schematic diagram of a concentric hemispherical electron energy analyser.

so that for a given incident flux density the peak to background ratio may be an order of magnitude greater than in AES. The counting time required to build up statistically significant data is correspondingly reduced.

Electrons are detected (counted) by the detector when they reach the fermi energy of the detector, and their kinetic energy is measured relative to the zero point of kinetic energy for the detector, and this is not necessarily the same as the zero point of kinetic energy for the initial source of electrons (the specimen). In order to measure the proper binding energy of electrons using XPS it is therefore essential to ensure that the sample and the detector are at the same potential (typically ground).

2.5. CHEMICAL ANALYSIS

2.5.1 Peak Positions
A survey spectrum covers a wide range of values of E_b, typically from $0\,eV$ to $1000\,eV$ or higher. The measured signals in E_{kin} would be converted to values of binding energy, and an ideal survey spectrum would appear as in Figure 2.4. Here it is assumed that the experiment is conducted with $T=0\,K$ with an ideal source and detector, and furthermore that Heisenberg's uncertainty principle does not operate, the electrons have no spin and that all the electrons created leave the sample with no losses.

Departure from this ideality leads to line broadening and to the appearance of additional lines in the spectra. For example, the presence of electrons with different spin-orbit coupling will have different binding energies resulting in splitting of the lines into two distinct energy levels (for all but s orbitals) whose total intensity

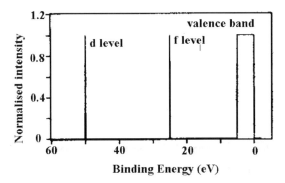

Figure 2.4. An ideal XPS survey spectrum: the core level peaks are of zero width.

(i.e. the sum of the two peak areas) is equal to the value it was without spin-orbit splitting.

The observed *peak width* ΔE (defined as the full width at half-maximum – FWHM) arises from several contributions, and may be expressed as:

$$\Delta E = (\Delta E_n^2 + \Delta E_p^2 + \Delta E_a^2)^{1/2}$$

where ΔE_n is the inherent width of the core level, ΔE_p is the width of the photon source and ΔE_a is the resolution of the analyser.

The *intensity* of a peak depends on the value of the photoelectron cross-section, σ, which is a measure of the efficiency of the photon interaction with the electron. Each orbital has its own cross-section, so the intensities of XPS peaks will not be identical even when all else is ideal.

It must also be remembered that electrons scatter from surrounding atoms as they exit the specimen, so that electrons will be observed with kinetic energies lower than that of the primary electron, giving rise to the presence of *loss peaks* in the spectrum. *Secondary electrons* (of low energy) will be ejected from the specimen by the primary electron, causing an increase in the background level of the spectrum. The Auger process (see p. 171) can occur when a hole is created in a core level, so that Auger peaks can also appear in XPS spectra.

The identification of the major peaks in the spectrum is accomplished by comparison with reference data (e.g. Wagner *et al.* 1978). The qualitative analysis of XPS spectra is more complex than for AES due to the presence of Auger peaks in addition to photoelectron peaks. If a photoelectron line of one element is close in energy to an Auger line of another, the problem may be resolved by taking spectra at two different photon energies.

Figure 2.5 illustrates two spectra recorded from a sample of iron using (a) Al K_α radiation, and (b) Mg K_α radiation. The binding energy of the peaks are characteristic of each element. There is a difference in hv between these sources of 233 eV, so, as expected from equation (2.1), the XPS peaks on spectrum (a) are displaced 233 eV relative to those in spectrum (b). The spectrum was taken over a wide energy range to detect all possible peaks of elements present in the surface. The 2p and 3p peaks from iron are identified, as well as the 1s peak from carbon which was present as a contaminant.

Additional peaks are present on the spectra arising from the ejection of Auger electrons (see p. 171). The Auger peaks result from electron binding energies alone (see p. 172), and are independent of the incident photon energy. So the LMM Auger peaks from iron, marked in Figure 2.5, are seen to be at fixed energies in spectra (a) and (b), and this permits their unambiguous identification.

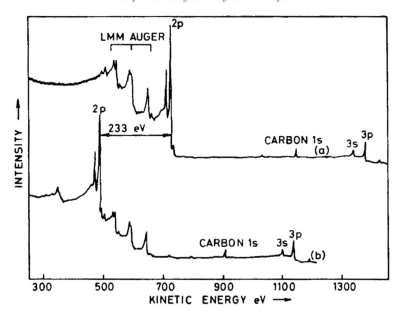

Figure 2.5. XPS spectra produced from a sample of iron which has been cleaned by argon ion bombardment, using (a) Aluminium K_α radiation, and (b) Magnesium K_α radiation. (Reproduced by permission of Flewitt and Wild 1985.)

2.5.2   *Chemical State Effects*

Special interest is shown in most cases not in the absolute value of E_b, but in its change ΔE_b for the electron level of the same element in different compounds. Any change in bonding of an atom that changes the binding energy of the electron of interest will cause a corresponding shift in the peak position. The value of ΔE_b is usually measured with respect to the free element, and is known as the *chemical shift*. An example of this would be a change in the state of oxidation, so Fe in FeO and Fe in Fe_2O_3 would have differing peak positions, since E_b of an electron in any orbital of an atom increases as the oxidation state of the atom increases.

Essentially, the stronger the chemical bonding experience by an ion, the more tightly bound its own energy levels become the higher the binding energy of the corresponding XPS peaks from those levels. In order to identify the chemical state of the element, accurate measures of the peak energies are therefore required, so that the energy scale of the spectrometer must be properly calibrated following recommended procedures and using known atomic standards for reference line energies.

2.5.3   *Quantitative Analysis*

For a *homogeneous binary sample* using reference samples in the same instrument:

$$\left(\frac{I_A}{I_A^\infty}\right)\left(\frac{I_B^\infty}{I_B}\right)=\left(\frac{\lambda_B(E_B)a_A^3}{\lambda_A(E_A)a_B^3}\right)\left(\frac{x_A}{x_B}\right) \tag{2.2}$$

where I_A, I_A^∞, I_B and I_B^∞ represent the signal from the material and pure reference sample for the elements A and B respectively, and E is the energy of the ejected electron in each case. λ_A and λ_B are the inelastic mean free paths in the solid, a_A and a_B are the atom sizes and x_A and x_B represent the concentrations of each constituent.

For most cases, equation (2.2) reduces to:

$$\left(\frac{x_A}{x_B}\right)=F^x_{AB}\left(\frac{I_A}{I_A^\infty}\right)\left(\frac{I_B^\infty}{I_B}\right) \tag{2.3}$$

where

$$F^x_{AB}=(a_B/a_A)^{3/2}$$

and is essentially a matrix factor which describes the difference in intensity expected in emission from atoms of element A embedded in pure A and atoms of element A in the substance AB.

Not all samples consist of binary mixtures, and difficulties exist with the extension of the matrix factor approach to multi-component systems.

If the composition of the outermost layer is different from that of the bulk (for example, as a result of surface or grain boundary segregation), a different approach has to be made. For example, if a partial overlayer of element A of fractional coverage φ_A covers a substrate of element B, the spectrum contains three contributions: that from the overlayer, that from the exposed part of the substrate, and that from the covered part of the substrate.

A monolayer matrix factor Q_{AB} can be defined such that:

$$\varphi_A=Q_{AB}\left(\frac{I_A/I_A^\infty}{I_B/I_B^\infty}\right) \tag{2.4}$$

where

$$Q_{AB}=\frac{\lambda_A(E_A)\cos\theta}{a_A}$$

θ is the angle between the sample normal and the spectrometer axis.

These equations can be expanded to incorporate a variety of more complex overlayer and substrate structures and compositions.

2.5.4 Spatial Resolution in XPS

The irradiating X-ray beam cannot be focussed upon and scanned across the specimen surface as is possible with an electron beam. Practical methods of small-spot XPS imaging rely on restriction of the source size or the analysed area. By using a focussing crystal monochromator for the X-rays, beam sizes of less than 10 μm may be achieved. This must in turn correspond with the acceptance area and alignment on the sample of the electron spectrometer, which involves the use of an electron lens of low aberration. The practically achievable spatial resolution is rarely better than 100 μm. A spatial resolution value of 200 μm might be regarded as typical, and it must also be remembered that areas of up to several millimetres in diameter can readily be analysed.

Even greater spatial resolution of the photoelectron output may be achieved by adjustments of the lens and slit system together with the application of focussing and rastering techniques (a resolution of 10 μm or less is obtained from these approaches). These recent improvements in technology have led to the development of *imaging XPS* equipment which either translate the sample position under the electron energy analyser, so that the analysed region is moved across the surface, or electrostatic deflection plates are incorporated within the electron optics to move the region from which electrons are collected across the sample surface.

The relatively poor spatial resolution of XPS compared, for example, with electron microscopy techniques such as SAM is more than offset by the benefit of concurrent chemical state identification.

2.6. AREAS OF XPS APPLICATION TO THE SURFACE ANALYSIS OF MATERIALS

Detailed and authoritative reviews of experimental techniques, and also of the areas of application of XPS are given by Barr (1994) and Briggs and Seah (1990). The applications of XPS include the study of the following.

2.6.1 The Natural Passivation and Corrosion of Metals and Alloys

XPS studies of the air-formed natural passive layer on aluminium surfaces have identified a number of hydroxides as well as alumina (Barr, 1977). The oxidation of pure iron and of stainless steels and other iron alloys have also been extensively investigated by this technique. XPS has been employed to identify corrosion inhibitors on oxide surfaces, as well as the nature of oxidation-resistant coatings in numerous systems.

2.6.1.1. Qualitative Analysis. A straightforward qualitative analysis of a surface film by XPS has been made in the case of the thermal oxidation behaviour of commercial aluminium alloys containing 2 to 3 weight percent lithium (Ahmad 1987). These have an attractive combination of mechanical properties – particularly in terms of reduced density and increased stiffness. Since lithium is highly reactive, one may expect that during thermomechanical processing of these alloys lithium may diffuse to the surface and react with the atmosphere. This can lead to the formation of a soft surface layer and thus a degradation in properties.

Ahmad studied specimens of an Al-Li-Cu-Mg-Zr alloy designated 8090 in the form of specimens between 1 mm and 3 mm in thickness cut from an extruded bar of cross-sectional dimensions 51 mm × 25 mm. An XPS spectrum of the surface of a sample oxidized for 5 min at 530°C in air is shown in Figure 2.6. In addition to the peaks of carbon, oxygen and magnesium, there is a lithium (Li 1s) peak at 56 eV binding energy.

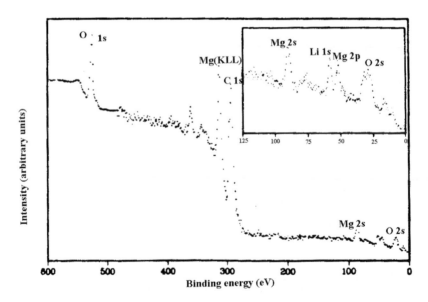

Figure 2.6. XPS spectra of the surface of a specimen of 8090 alloy heat treated for 5 min at 530°C in laboratory air. (After Ahmad 1987.)

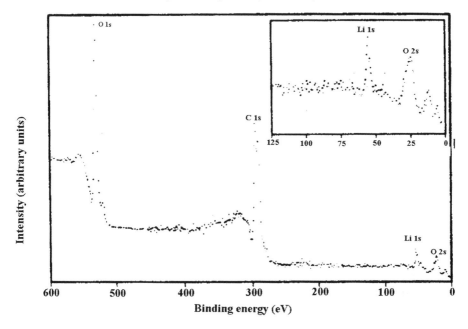

Figure 2.7. XPS spectra of the surface of 8090 alloy heat-treated for 30 min at 530°C in laboratory air. (After Ahmad, 1987.)

The spectra obtained from the surface of a specimen oxidized 30 min are shown in Figure 2.7. There are peaks of carbon, oxygen and lithium, while the peaks of magnesium are absent.

These observations (confirmed by AES studies) indicate that after longer oxidation times the top surface is completely covered by lithium compounds, i.e. that the oxidation of lithium has become dominant. This implies a depletion of the element within the metal surface, i.e. the presence of a soft surface layer.

2.6.2 *The Composition Profile of Surface Films on Metals*

Jin and Atrens (1987) have elucidated the structure of the passive film formed on stainless steels during immersion in 0.1 M NaCl solution for various immersion times, employing XPS and ion etching techniques. The measured spectra consist of composite peaks produced by electrons of slightly different energy if the element is in several different chemical states. Peak deconvolution (which is a non-trivial problem) has to be conducted, and these authors used a manual procedure based on the actual individual peaks shapes and peak positions as recorded by Wagner *et al.* (1978). The procedure is illustrated in Figure 2.8 for iron.

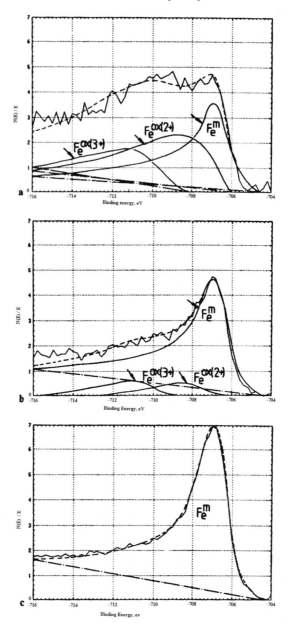

Figure 2.8. (a–c) Deconvolution of Fe peaks into their constituents in terms of metallic iron, Fe^m, Fe^{2+}, and Fe^{3+}. The dashed curve, the sum of the individual constituent curves, is in good agreement with the measured curve. Also shown are the assumed backgrounds. The specimen is an 18–12 austenitic stainless steel which, after air exposure, spent 44 h in 0.1 M NaCl solution. The sputter time was (a) 30 s; (b) 60 s, and (c) 15 min. (After Jin and Atrens, 1987.)

The structure and composition of the passive films were determined using the depth profiling technique, consisting of periodic sputtering with argon ions and recording the electron spectra after each etching period. The sputtering rate was related to that measured using a standard specimen, and Figure 2.9a–d show the actual composition profiles for the steel after different times of immersion in the NaCl solution.

The curves of Figure 2.9 exhibit the complex structure of the surface film. With increasing depth there is a peak of iron in the oxidized state at approximately 0.3 nm, and a peak of chromium in the oxidized state at about 1 nm irrespective of immersion time. The maximum concentration of oxidized iron decreases and the maximum concentration of oxidized chromium increases with increasing immersion time.

2.6.3 Glasses

A wide range of problems have been studied, including chemical stability, the strengthening of glass, and the distribution of tin in the surface of float glass.

Glasses present an experimental problem common to the study of all electrical insulators, namely that since electrons are emitted from the sample during analysis, the surface of the specimen will be charged positively. The surface charging leads to problems of instability making analysis impossible, and a shifting of the spectrum on the energy scale, leading to difficulty in interpreting chemical states. Use of an electron 'flood gun' technique may overcome these problems. This consists of a heated filament above the specimen which is held at a negative voltage relative to the sample, thus providing a low-energy electron flux over the sample surface. Figure 2.10 gives an illustration of electron flood gun treatment of XPS charging shifts encountered with a specimen of SiO_2.

2.6.4 Semiconductors

The surface oxidation of semiconducting materials is of both scientific and practical importance, and XPS has contributed to solving questions concerning the composition of the oxidised layer and the first stages of oxygen absorption in semiconductors. For example, the preferential adsorption of oxygen on the surface of GaAs has been studied, revealing that during heating there is a transfer of oxygen from the As–O bond to the Ga–O bonds.

2.6.5 Polymer Technology

This general area has been reviewed in depth by Briggs (1990), who gives an extended bibliography of the application of XPS to a number of aspects, including plasma treatment, photooxidation and weathering, biomedical polymers and polymer–metal interactions.

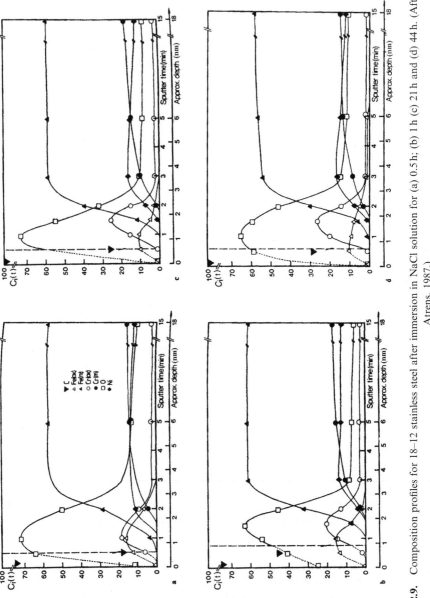

Figure 2.9.   Composition profiles for 18–12 stainless steel after immersion in NaCl solution for (a) 0.5 h; (b) 1 h (c) 21 h and (d) 44 h. (After Jin and Atrens, 1987.)

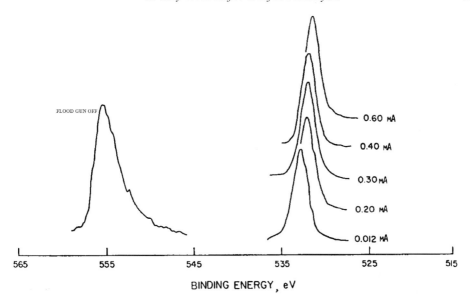

Figure 2.10. An example of electron flood gun treatment of XPS charging shifts of O (1s) lines from a silica specimen. The effect of increasing flood gun current is shown. (After Barr, 1983.)

The common polymers are composed of a small number of elements whose XP spectra are simple (generally C 1s plus one or two peaks from O 1s, N 1s, F 1s and Cl 2s, 2p). Common contaminants contain additional elements such as S, P, Si, Al and heavy metals, and the presence of these elements, even in low concentrations, can be detected very easily. Polymer surface modification is an area in which XPS has been fruitfully applied, notably in the study of commercial pretreatments aimed at improving wettability and general adhesion characteristics.

Problems of specimen charging have again to be considered.

2.6.6 Catalysis

Zeolites, clays, and platinum metal catalysis. This area is reviewed in detail by Barr (1990), and here again problems of specimen charging may arise.

2.6.7 Fracture Surfaces of Metals and Alloys

Auger spectroscopy (q.v.) is usually employed in the study of cleavage surfaces, but an advantage of XPS in such studies is that the chemical nature of elements segregating on the surface can be investigated.

REFERENCES

Ahmad, M. (1987) *Metallurgical Transactions*, **18A**, 681–689.

Barr, T.L. (1977) *J. Vac. Sci. Technol.*, **14**, 660.

Barr, T.L. (1990) In Applications of electron spectroscopy to Heterogeneous Catalysis in Briggs, D. and Seah, M.P. (eds.) *Practical Surface Analysis*, 2nd edn., John Wiley & Sons, Chichester, England.

Barr, T.L. (1994) *Modern ESCA – The Principles and Practice of X-ray Photoelectron Spectroscopy*, Boca Raton, Florida, CRC Press.

Briggs, D. (1990) *Applications of XPS in Polymer Technology* in Briggs, D. and Seah, M.P. (eds.) (1990) *Practical Surface Analysis*, 2nd ed., John Wiley & Sons, Chichester, England.

Briggs, D. & Seah, M.P. (eds.) (1990) *Practical Surface Analysis*, 2nd ed., John Wiley & Sons, Chichester, England.

Flewitt, P.E.J. & Wild, R.K. (1985) *Microstructural Characterisation of Metals and Alloys*, London, The Institute of Metals.

Jin, S. & Atrens, A. (1987) *Appl. Phys.*, **A42,** 149–165.

ASTM (1997) Standard Guide for Procedures for Specimen Preparation and Mounting in Surface Analysis E1078-97.

Smith, G.C. (1991) *Quantitative Surface Analysis for Materials Science*, London, The Institute of Metals.

Wagner C.D., Riggs W.M., Davis L.E., Moulder G.E. & Muilenberg G.E. (1978) *Handbook of X-ray Photoelectron Spectroscopy* Perkin-Elmer Corporation, Eden Prairie.

FURTHER READING

Barr, T.L. (1983) *Appl. Surface Science*, **15**, 1–35.

Nefedov, V.I. (1988) *X-ray Photoelectron Spectroscopy* VSP BV, Utrecht, The Netherlands.

Chapter 3

Infrared (IR) and Ultraviolet (UV) Probes for Surface Analysis

Chapter 3
Infrared (IR) and Ultraviolet (UV) Probes for Surface Analysis

3.1. IR SPECTROSCOPY

Vibrational spectroscopy provides the most definitive means of identifying the surface species arising from molecular adsorption and the species generated by surface reaction, and the two techniques that are routinely used for vibrational studies of molecules on surfaces are Infrared (IR) Spectroscopy and Electron Energy Loss Spectroscopy (HREELS) (q.v.).

IR spectroscopy is a highly versatile technique, being applicable to almost any surface, and capable of operating under both high and low pressure. It is of relatively low cost compared to a technique which requires high vacuum for operation. There are a number of ways in which IR techniques may be employed in the study of adsorbates on surfaces. A great advantage of infrared spectroscopy is that the technique can be used to study catalysts *in situ*. For solids with *a high surface area* these are:

Transmission IR (TIR) spectroscopy if the solid in question is IR transparent over an appreciable range of wavelength. This is often used on supported metal catalysts, where the large metallic surface area permits a high concentration of adsorbed species to be sampled. The sample consist typically of 10–100 mg of catalyst, pressed into a self-supporting disk of approximately $1 \, cm^2$ and a few tenths of a mm in thickness. The support particles should be smaller than the wavelength of the IR radiation, otherwise scattering losses become important.

Diffuse Reflectance IR Fourier Transform Spectroscopy (DRIFTS) can be employed with high surface area catalytic samples that are not sufficiently transparent to be studied in transmission. In this technique, the diffusely scattered IR radiation from a sample is collected, refocussed, and analysed. Samples can be measured in the form of loose powders.

For solid samples with *a low surface area* one may employ:

Reflection-Absorption IR spectroscopy (RAIRS) where the linearly polarized IR beam is specularly reflected from the front face of a highly reflective sample, such as a metal single crystal surface (Figure 3.1(a)). This is also sometimes referred to as *IRAS* (IR reflection absorption). The IR beam comes in at grazing angle (i.e. almost parallel to the surface), and although absorption bands in RAIRS have intensities that are some two orders of magnitude weaker than in transmission studies on

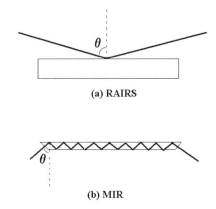

(a) RAIRS

(b) MIR

Figure 3.1.   Showing geometries for (a) RAIRS (IRAS) , where $80° < \theta < 89°$ and (b) Multiple Internal
Reflection spectroscopy (MIR), where $\theta > \theta_{critical}$.

supported catalysts, RAIRS spectra can be measured accurately with standard
spectrometers. Chabal (1988) gives a review of the technique.

Multiple Internal Reflection Spectroscopy (MIR) is an alternative approach in which
the IR beam is passed through a thin, IR transmitting sample such that it undergoes
total internal reflection alternately from the front and rear faces of the sample
(Figure 3.1(b)). At each reflection, some of the IT radiation may be absorbed by
species adsorbed on the solid surface.

Infrared radiation falls into three categories, as indicated below:

Region	Wavelength (μm)	Energy (meV)	Wave numbers (= frequency/c)
Far IR	1000–50	1.2–25	10–200
Mid IR	50–2.5	25–496	200–4000
Near IR	2.5–1	496–1240	4000–10,000

It is the mid-infrared region that is used for the study of molecular vibrations and is
therefore of greatest importance in the present context. The first studies in the mid-
IR on catalysis were by Eischens and Pliskin (1958) whose review is still a standard
work of reference.

3.1.1   Theory of Molecular Vibrations
For IR spectroscopy, the process of interest is absorption. Polarization and angle-
dependent measurements are useful when using the transmission geometry.

In general, when using unpolarized light, the vibrations both parallel and perpendicular to the interface are excited. However, when the light is polarized, selected dynamic dipoles are excited, from which the orientation of the dipoles can be inferred.

The maximum sensitivity in the study of adsorbates on metallic surfaces is obtained using the grazing incidence reflection methods (RAIRS). The observation of vibrational modes of such adsorbates is subject to the *surface dipole selection rule*. Dynamic dipoles with displacements perpendicular to the interface are excited by the polarization component in the incident beam that is perpendicular to the interface (p-polarized), and the excitations are called the *dipole-allowed* modes. These take place when the frequencies of the light and the oscillations are similar (resonance). The polarization component of the beam parallel to the surface (s-polarized) destructively interferes with itself upon reflection, and cannot directly excite vibrations with displacements parallel to the surface. Thus only those vibrational modes which give rise to an oscillating dipole perpendicular to the surface are IR active and give rise to an observable absorption band. The optimal angle of incidence is dependent upon the conductivity of the reflective material, and the energy of the excitation.

Molecules possess discrete levels of rotational and vibrational energy, and transitions between vibrational levels occur by absorption of photons with frequencies in the mid-infrared range. There are four types of vibration:

Stretch vibrations (ν) changing the length of the bond
Bending vibrations (δ) changing bond angles but leaving bond lengths unaltered.
Bending vibrations out of plane (γ)
Torsion vibrations (τ) changing the angle between two planes through atoms.

The frequencies of these vibrations generally decrease in the order $\nu > \delta > \gamma > \tau$. Not all vibrations can be observed: absorption of an IR photon occurs only if a dipole moment changes during the vibration. The intensity of the IR band is proportional to the change in dipole moment. Thus species with polar bonds (e.g. CO, NO and OH) exhibit strong IR bands, whereas molecules such as H_2 and N_2 are not infrared active at all.

An absorption spectrum is a plot that shows how well different frequencies of light couple to excitations in the sample. It is conventional to convert the units for frequency (ν) from Hertz to wave numbers (cm^{-1}) by dividing ν by the speed of light (c). IR frequencies are characteristic of certain bonds in molecules and can thus be used to identify species on surfaces. Correlation charts are available which permit assignments of particular molecular species to certain IR frequencies.

3.1.2 Equipment

Most modern IR facilities will use a Fourier Transform IR Spectrometer (FTIR), rather than a dispersive instrument. The essential feature is that all of the light from the source falls on to the detector at any instant, which thus leads to increased signal levels, thereby automatically improving the signal-to-noise ratio at all points on the spectrum.

One major problem is that of sensitivity (i.e. the signal is very weak owing to the small number of adsorbing molecules). Typically the sampled area is $\sim 1\,cm^2$ with less than 10^{15} adsorbed molecules (i.e. about 1 nmol). With modern FTIR spectrometers, however, such small signals (0.01–2% absorption) can still be recorded at relatively high resolution ($\sim 1\,cm^{-1}$).

The source is usually a temperature-stabilized ceramic filament operating around 1500 K. The detector in FTIR is usually a deuterium triglycine sulphate (DTGS) detector, although in RAIRS experiments the liquid nitrogen-cooled mercury cadmium telluride (MCT) detector is employed.

Two equally intense beams from the source are produced optically by a simple beam-splitter device, each beam containing all wavelengths emanating from the source. Frequency analysis (Fourier analysis) takes place at the detector where a Michelson interferometer (or similar device) induces a periodically varying path length difference between the two beams. The two beams are recombined and detected: the intensity measured depends on the overall effects of phase difference for each component wavelength. The Fourier transformation is the mathematical operation which converts the signal which varies with path length to a spectrum in which intensity varies with wavelength. These instruments have the advantage that the entire spectrum is obtained for each scan the interferometer makes, so that the total collection time needed to measure a spectrum is much lower than in the case of an energy-dispersive type of spectrometer which selects for analysis a single wavelength of interest. Ferraro and Basilo (1978, 1979), for example, give further information on the operation of an FTIR instrument.

Hirschmugl (2001) gives an overview of recent advances in IR spectroscopy at surfaces and interfaces, considering not only the advances in experimental methods but also the future directions in which the technique may be applied.

3.1.3 Some Applications

3.1.3.1 Catalysis. Beitel *et al.* (1997) have employed RAIRS to study *in situ* the co-adsorption behaviour of CO and hydrogen on single-crystal cobalt (0001) catalysts at pressures up to 300 mbar and temperatures between 298 and 490 K. The behaviour of these adsorbates is of considerable importance in relation to their commercial importance as catalysts for the Fischer-Tropsch reaction in the manufacture of long-chain hydrocarbon products, and detailed understanding of the co-adsorption behaviour has been hampered by the lack of available *in situ* techniques. Conclusions drawn from results obtained under UHV (ultrahigh vacuum) conditions are of unknown relevance to industrial catalytic practice, where temperatures between 450 and 570 K are employed at pressured typically in the range 5–30 bar.

In applying RAIRS to CO adsorption, the contribution from CO molecules in the gas phase to the absorption spectrum at CO pressures above 10^{-3} mbar completely obscures the weak absorption signal of surface adsorbed CO. Beitel *et al.* found it possible to subtract out the gas phase absorption by coding the surface absorption signal by means of the polarization modulation (PM) technique applied to a conventional RAIRS spectrometer. p-polarised light produces a net surface electric field which can interact with adsorbed molecules, whereas both polarization states are equally sensitive to gas phase absorption because gas phase molecules are randomly oriented. By electronic filtering a differential spectrum is computed which does not show contributions from the gas phase and which has much higher surface sensitivity than a conventional RAIRS setup.

The catalytic preformance of Co crystals with two surface conditions were compared: annealed crystals with large atomically flat terraces and Ar^+ ion sputtered surfaces which produced a high population of surface defects. A sequence of PM-RAIRS spectra are shown in Figure 3.2 during exposure of a sputtered Co (0001) surface to mixtures of H_2 and CO, with the temperature and pressure for each spectrum indicated in the figure.

Figure 3.2 shows that at 298 K the spectra under pure CO and under synthesis gas (100 mbar CO + 200 mbar H_2) are closely similar. The rest of the sequence shown in Figure 3.2 shows the effect of raising the temperature stepwise to 490 K under synthesis gas. Each spectrum was obtained after holding the sample for 5 min at the indicated temperature. The sharp defect peak at 2080 cm^{-1} shows a large decrease in intensity with increasing temperature, accompanied by a small shift to lower frequencies. The peak at $\sim$2020–2060 cm^{-1} shows an increase in intensity with increasing temperature and is accompanied by a shift to lower frequencies. Lowering the temperature again to 298 K results in the shift of the terrace absorption signal to higher wavenumbers, although the 2080 cm^{-1} absorption signal does not develop again, clearly showing the irreversibility of this process.

The interpretation of the IR data and subsequent X-ray photoelectron spectros-copy studies in UHV conditions indicate that at room temperature CO is adsorbed on the Co surface. Under higher temperatures, H starts to occupy some available sites, and hydrocarbons form at the defect sites. Finally, at room temperature and under vacuum conditions the hydrogen desorbs from the Co, while the Co and hydrocarbons remain.

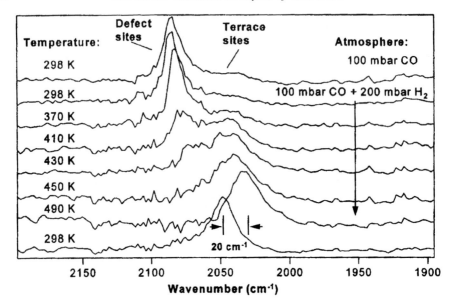

Figure 3.2. Sequence of PM-RAIR spectra taken on a sputtered Co(0001) surface. The starting conditions were 100 mbar of CO at 298 K; 200 mbar of H_2 was then added and the temperature increased stepwise to 490 K. It can be seen that the absorption signal due to CO attached to defect sites is removed in an irreversible process. The lowest curve was obtained at room temperature and under vacuum conditions and the hydrogen desorbs from the Co, while the CO and hydrocarbons remain. (After Beitel *et al.* 1997.)

3.1.3.2 *Electrochemical Cells.*

IR spectroscopy provides an *in situ* probe of the constituents adsorbed at electrode surfaces, thus proving to be a valuable tool for understanding reaction pathways in these complex environments. IR spectra acquired at different points in a voltammogram can be compared, and this may elucidate the electrochemistry of a given process.

As an example of this application we will examine the work of Erné *et al.* (1998), who studied the dynamics of hydrogen adsorption on *n*-type GaAs (100) electrodes. The cathodic production of hydrogen gas is important in the context of its promise as a fuel for the chemical storage of energy. In this reaction, two electrons are transferred from the electrode to two aqueously solvated protons H^+(aq), yielding a hydrogen gas molecule H_2 (g); a hydrogen atom is temporarily adsorbed after transfer of the first electron. The spectroscopic detection of adsorbed hydrogen is based on the resonant vibrational absorption of hydrogen-to-surface bonds.

Surface hydrogen is conveniently detected by MIR at attenuated-total-internal-reflection prisms of GaAs electrodes. Si-doped ⟨100⟩-oriented *n*-GaAs single crystals were employed in the electrochemical cell illustrated schematically in Figure 3.3.

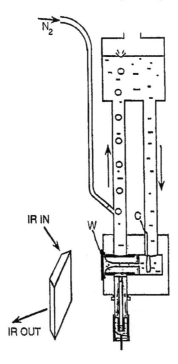

Figure 3.3. Schematic diagram of the electrochemical cell: inset shows the sample geometry for MIR experiments (After da Fonseca *et al.* 1996.)

The vertically mounted GaAs electrode (45° angles), W (Figure 3.3) exposed an area of $1\,cm^2$ to a circulating aqueous solution of HCl or NaOH, a mild nitrogen bubbling being used as a circulating pump. The electrodes, of typical dimension $13 \times 15 \times 0.5\,mm^3$, were pressed with an O-ring seal against the cell wall. Uniform accessibility of the electrode was approached by shaping the PTFCE pipe in front of the electrode as a tuyère. The apparatus included a platinum counter electrode and an Ag/AgCl in saturated KCl reference electrode.

IR absorbance was measured with a Fourier-transform IR spectrometer. The absorbance at wave number σ is defined as $(1/N)\ln[I^o(U_o)/I^o(U)]$, where $N \approx 10$ is the number of useful reflections at the electrochemical interface, $I^o(U)$ the light intensity at wave number σ reaching the detector at potential U, and $I^o(U_o)$ the same but under reference conditions at potential U_o.

Hydrogen binds to As sites at the surface when the GaAs electrode is electron rich; when the GaAs electrode is electron poor, the hydrogen adsorbates are replaced by OH species at the As sites. Changes in potential were determined by interrupting the cyclic potential scans every 100 ms for a 1 mn period at various

positive potentials, at which the capacitance of the semiconductor space charge layer
was measured. Figure 3.4 shows the change in interfacial IR absorbance as a result
of going from the potential where no current is measured in the steady state (-0.3 V
vs. Ag/AgCl) to a potential where hydrogen evolution occurs (-0.8 V vs. Ag/AgCl).

It can be seen in Figure 3.4 that absorbance increases at $2050\,\text{cm}^{-1}$ (which
corresponds to As–H bands only) and decreases in the 2500–$3700\,\text{cm}^{-1}$ range (which
is ascribed to loss of surface OH-groups present at the surface in the anodic range
and replaced by adsorbed hydrogen in the cathodic range). The As–H bonds were
found to absorb p-polarised light more strongly than s-polarised light, confirming
that they are at the surface rather than inside the solid. It should be noted that these
measurements are sensitive only to *changes* at the interface, and therefore are not
sensitive to adsorbate occupation that does not change over the voltammogram
cycle.

3.1.3.3   Interface Structure of Devices.   IR spectroscopy can provide structural
information about technologically important oxide/metal and oxide/semiconductor
interfaces. For example, Stefanov *et al.* (1998) were able to identify the intermediate
oxide structures formed by high temperature annealing of a water-exposed Si(100)
surface. Employing FTIR spectrometry in conjunction with transmission geometry
($60°$ incidence to the surface normal) increased the sensitivity to a wider spectral
range. This allowed the simultaneous observation of the hitherto unobserved
low frequency Si–O stretching and Si–H bending fundamentals as well as the

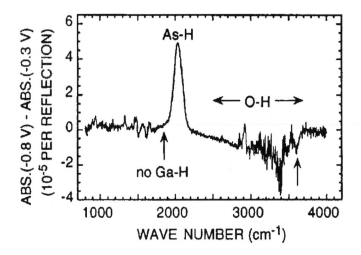

Figure 3.4.   Change in IR absorbance of *n*-GaAs/ 6 M HCl interface when the applied potential is
changed from -0.3 to -0.8 vs. Ag/AgCl (1500 cycles of 12 s). (After Erné *et al.* 1998.)

higher-frequency Si–H and O–H stretching modes. These data, in conjunction with first principles theoretical calculations have identified that the SiO_2/Si interface contains a Si_2O_3 and a Si_2O_5 species.

3.1.3.4 Biocompatibility.

3.1.3.4 Biocompatibility. The analysis of polymer implants has been employed using FTIR spectroscopy to elucidate the long-term biocompatibility and quality control of biomedical materials. This method of surface analysis allows the determination of the specific molecular composition and structures most appropriate for long-term compatibility in humans.

Afanasyeva and Bruch (1999) have applied the FTIP reflection method to investigate the surfaces of materials used for intraocular lenses (IOLs) in the production of implants that are non-toxic to the eye. In particular they studied passivated lens surfaces, which is a treatment which impedes the growth of cells upon the lens surface. Lenses made of PMMA and of sapphire were examined, and Figure 3.5 shows a typical IR spectrum in the 3000–2800 cm^{-1} wavenumber region.

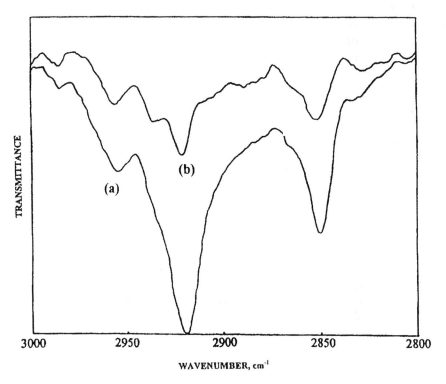

Figure 3.5. Typical FTIR spectra for sapphire intraocular lenses: (a) surface not passivated; (b) surface is passivated. (Reproduced by permission of Afanasyeva and Bruch 1999.)

It is evident in Figure 3.5 that the two displayed spectra are slightly different in the band intensities and observed spectral features. This approach is thus suitable for analysing the characteristic band structures to enhance the bio-compatibility of the sapphire lenses, and the surface passivation process enabled more optimized biocompatible lenses to be fabricated.

3.2. RAMAN MICROSCOPY

Infra-red (IR) spectroscopy (q.v.) requires a change in dipole moment when a molecule is excited by incident energy in the form of photons, in contrast to Raman spectroscopy, which requires a change in the *polarizability* of an excited vibrational state of a molecule. Raman scattering was discovered by the Indian physicist C.V. Raman in 1928, but it was not applied to chemical analysis to any significant extent until after 1986 with the introduction of Fourier transform (FT)-Raman charge-coupled devices, small computers and near-infrared lasers. These technological developments largely overcame the fundamental problems of a weak Raman signal and interference from fluorescence.

FTIR is the most widely used vibrational spectroscopy, but it does have some limitations that are fundamental to the wavelength range involved. Mid-IR light does not penetrate many common optical materials such as glass, thus restricting sampling flexibility. Near infrared (NIR) absorption uses shorter wavelength light (1–2.5 µm) than FTIR and is compatible with common glass, although NIR absorptions are both weaker and broader than FTIR bands. The attraction of Raman spectroscopy is that it combines many of the advantages of FTIR with those of NIR absorption.

3.2.1 Principle

A theoretical account of Raman scattering is given by Turrell (1996) and by McCreery (2000), to which the reader is referred for a fuller treatment than that which is given below. Photons interact with molecules to induce transitions between energy states. Most photons are elastically scattered (Rayleigh scattering) so the emitted photon has the same wavelength as the absorbing photon. Raman spectroscopy involves the inelastic scattering of photons by molecules, as indicated in the simplified diagram of Figure 3.6.

The energy of the scattered radiation is less than that of the incident radiation for the *Stokes line* of the Raman spectrum and the energy of the scattered radiation is more than that of the incident radiation for the *anti-Stokes line*. The energy increase or decrease from the excitation is related to the vibrational energy spacing

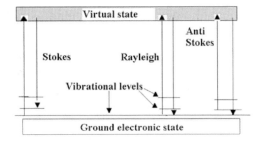

Figure 3.6. A simplified energy diagram illustrating the origins of Rayleigh scattering and of the Stokes and anti-Stokes lines in the Raman spectrum.

in the ground electronic state of the molecule and therefore the wavenumbers of the Stokes and anti-Stokes lines are a direct measure of the vibrational energies of the molecule.

The electromagnetic field of the incident radiation, E_i, may induce an electric dipole in the molecule, given by:

$$p = \alpha E_i$$

where α is the *polarizability* of the molecule and p is the induced dipole. The electric field due to the incident radiation is a time-varying quantity of the form:

$$E_i = E_0 \cos(2\pi v_i t) \tag{3.1}$$

For a vibrating molecule, the polarizability is also a time-varying term that depends on the vibrational frequency of the molecule, v_{vib}:

$$\alpha = \alpha_0 + \alpha_{vib} \cos(2\pi v_i t) \tag{3.2}$$

Multiplication of (1) and (2) gives rise to a cross-product term of the form:

$$\alpha_{vib} E_0/2[\cos 2\pi t(v_i + v_{vib}) + \cos 2\pi t(v_i - v_{vib})]$$

which represents light that can be scattered at both higher and lower energy than the Rayleigh scattering of the incident radiation, i.e. the anti-Stokes and Stokes lines respectively.

A Raman spectrum consists of scattered intensity plotted versus energy: each peak corresponds to a given Raman shift from the incident light energy. If the molecule happens to be in an excited vibrational state when an incident photon is scattered, the photon may gain energy when scattered, leading to anti-Stokes scattering. The Stokes and anti-Stokes peaks are symmetrically positioned about the Rayleigh peak, but the anti-Stokes line is much less intense than the Stokes line. In Raman spectroscopy, only the more intense Stokes lines are normally measured.

Since the Raman scattering is not very efficient (only one photon in 10^7 gives rise to the Raman effect), a high power excitation source such as a laser is needed. Also, since we are interested in the energy (wavenumber) difference between the excitation and the Stokes lines, the excitation source should be monochromatic, which is another property of many laser systems.

In crystalline solids, the Raman effect deals with phonons instead of molecular vibration, and it depends upon the crystal symmetry whether a phonon is Raman active or not. For each class of crystal symmetry it is possible to calculate which phonons are Raman active for a given direction of the incident and scattered light with respect to the crystallographic axes of the specimen. A table has been derived (Loudon, 1964, 1965) which presents the form of the scattering tensor for each of the 32 crystal classes, which is particularly useful in the interpretation of the Raman spectra of crystalline samples.

3.2.2 Apparatus

Minimal sample preparation is involved in Raman spectroscopy: samples as thin as a monolayer can be examined.

Visible lasers are typically used for sample excitation, although near-IR lasers can be used when visible excitation sources cause sample fluorescence, obscuring the Raman scatter.

A variety of optical systems have been described which use optical fibres to perform remote Raman spectroscopic analyses. These are of particular value when the sample has to be observed under hostile or hazardous environmental conditions, such as the in-line monitoring of chemical production. The collection radius of such fibre-optic probes is of the order 100 μm.

The spectroscopy system uses a dispersive element and a detector which is either a charge-coupled device (CCD) or a diode array. A computer is required for instrument control and for intensive data processing.

The sample spectrum is compared to a database of known spectra. For example, in the study of polymers, the most prominent regions are the fingerprint region (1200–$3000\,\text{cm}^{-1}$) and the 500–$1200\,\text{cm}^{-1}$ region that shows some response for characteristic functional groups. The intensity of Raman scattering by a functional group is proportional to its concentration, so that the relative concentrations of organic compounds which exist in several isomeric forms may be readily determined. For example, polybutadiene has three isomeric forms, and their relative concentrations is an important parameter which governs the mechanical properties of its copolymer with polystyrene (Dhamelincourt and Nakashima, in Turrell and Corset 1996). In the 1600–1700 cm^{-1} spectral range the C=C stretching modes of the three isomers appear separately in the Raman spectrum, so that the ratio of these isomers may be determined *in situ*, and microscopic inhomogenities can thus be detected.

When detailed information is needed about local variations in composition, *Raman microscopy* is used.

3.2.3 *Raman Microscopy and Imaging*

Raman microscopy has been used for analysis of very small samples or small heterogeneities in larger samples. Recent developments and applications of this technique have been reviewed by Turrell and Corset (1996), including a discussion of the coupling of Raman microscopy with electron, ion and x-ray microscopies, and these authors give a description of a number of prototype instruments with this facility.

A schematic diagram of conventional Raman microscopy, due to Turrell and Corset (1966), is given in Figure 3.7 which illustrates the laser focussing, sample

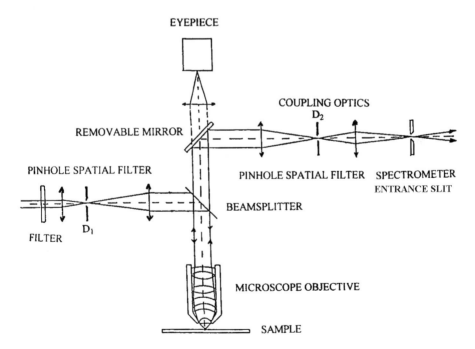

Figure 3.7. Schematic diagram of the basic layout of the apparatus typically employed in micro-Raman spectrometers, or microprobes. (From Turrell and Corset 1996.)

viewing and scattered light collection geometry which is widely employed in micro-Raman spectrometers.

Figure 3.7 shows an illumination pinhole, D_1, which filters the laser excitation beam to give a clean point source which is imaged on the sample. The scattered Raman radiation is collected by a wide-aperture objective and focused on an adjustable pinhole D_2 which is in the image plane of the microscope. D_1 and D_2 are known as *confocal* diaphragms, and they permit 'optical sectioning' of the sample by controlling the axial resolving power (or depth of focus) of the instrument. Decreasing the size of the confocal diaphragm D_1 improves both the axial resolution and reflection of stray light. Thus when thin or heterogeneous samples are examined, the diaphragm is adjusted so that only the Raman flux originating from the interesting part of the sample is transmitted to the detector.

Several imaging methods may be employed in Raman microscopy. Heterogeneity in a sample may be explored by establishing a *point-to-point* profile, which is a series of single-point acquisitions. The complete spectrum of each point is stored for subsequent analysis. In *line imaging*, the laser is focused to a line, then the scattering is projected to a CCD to generate a two-dimensional data set of intensity vs. Raman shift and position along the line. *Global imaging* illuminates a relatively large sample area, but spatially resolved spectra are collected. For example, a filter that transmits only a particular Raman shift might precede the CCD, providing a chemically selective two-dimensional image.

3.2.4   Applications of the Raman Microprobe

The microprobe has the ability to investigate regions as small as 1 μm in diameter, yielding molecular information, and it may complement other microanalysis techniques. It is a unique technique for the investigation of organic substances, and examples will be selected below which illustrate what Raman microspectroscopy can offer in the investigation of both organic and inorganic materials.

We will confine ourselves to those applications concerned with chemical analysis, although the Raman microprobe also enables the stress and strain imposed in a sample to be examined. Externally applied stress-induced changes in intramolecular distances of the lattice structures are reflected in changes in the Raman spectrum, so that the technique may be used, for example, to study the local stresses and strains in polymer fibre and ceramic fibre composite materials.

3.2.4.1   Inorganic Compounds.

The Raman microprobe has been used by Berg and Kerridge (2000) to obtain computer mappings of the structures of salt eutectics solidified from their melts. In contrast to metallic eutectics, which commonly occur

Table 3.1.

Eutectic mixture	Composition (mol%)	Melting point ($^{\circ}$C)
$NaNO_2/NaNO_3$	59:41	227
KNO_2/KNO_3	24:76	316
$NaNO_3/KNO_3$	50:50	220
$KNO_3/Ca(NO_3)_2$	65.8:34.2	145
$NaNO_2/NaNO_3/KNO_3$	48.9:6.9:44.2	142
KCl/K_2CrO_4	25:75	368

in the form of rapidly alternating phases of lamellar or fibrous forms, the components of the salt eutectics were concentrated into roughly rounded areas of about 0.5 to 5 µm across, which the authors describe as 'conglomerate'.

The eutectics examined and their melting points are given in Table 3.1. The compounds were melted in 10 mm inner diameter pyrex tubes, and polished sections were prepared from the solidified eutectic samples.

Small samples were placed on the microscope table. This table was operated from a computer that also controlled the data acquisition.

Raman spectra of the investigated ions are well known, and the sample spectra were excited with an Ar^+ ion laser light of about 1 W power at 514.5 nm wavelength. The Rayleigh line was filtered off and the Raman light, and the Raman light was dispersed into a liquid-nitrogen cooled CCD detector. The slits were set to 100 µm corresponding to a spectral resolution of about $4-5\,cm^{-1}$. Calibration of wavenumber scales to an accuracy of $\pm 1\,cm^{-1}$ was achieved by superimposing neon lines on the spectra. The spot size was generally 1 µm and a 40×40 matrix of points was normally used.

Figure 3.8 illustrates the type of map obtained: the spatial resolution was 1 µm, obtained with a 100× objective, and a 180 µm confocal aperture and a 100 µm slit width. Under the optical microscope the polished surfaces of this and the other samples were virtually featureless. The same segregation features were seen for other Raman mappings of this and the other systems, and was independent of the plane of section. The chemical components are seen to be gathered into irregular masses of about 0.5 to 5 µm across, with compositions ranging from high in nitrite to high in nitrate.

The use of Raman microscopy in the detection and identification of pigments on manuscripts, paintings, ceramics and papyri was reviewed by Clark (1999). He concludes that it is arguable the best single technique to be applied to this area, since it combines the attributes of reproducibility and sensitivity with those of being non-destructive and immune to interference from both pigments and binders. He points

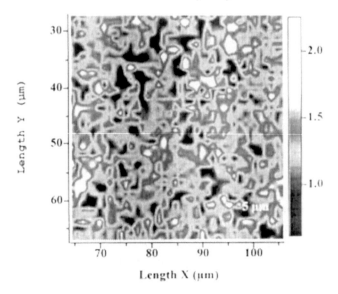

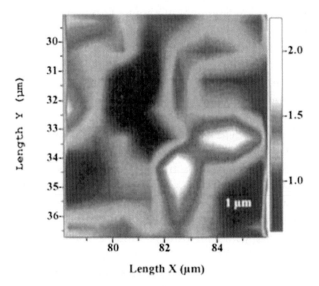

Figure 3.8. (a) Raman mapping of a 59/41 molar ratio sample of $NaNO_2/NaNO_3$ eutectic solidified at a cooling rate of $0.1°C/hr$. The solid was sectioned at $90°$ to the tube axis. The mapping is based on 40×40 spectra providing the intensity ratio for the bands seen at $\sim 1329 \, cm^{-1}$ ($NaNO_2$) and $\sim 1068 \, cm^{-1}$ ($NaNO_3$). (b) is an enlargement of a section of the mapping. (Berg and Kerridge 2000.)

out that the technique can be applied *in situ* (which makes it important for the study of manuscripts for which sampling is forbidden), and it has high spatial (≤ 1 μm) and spectral (~ 1 cm^{-1}) resolution.

A library of Raman spectra of c. 60 pigments, both natural and synthetic, known to have been used before 1850 AD has been compiled and published (Clark and Gibbs 1997). Two important technical developments involve the use of a fibre-optic probe, which makes possible the study of pigments on wall paintings, statues, etc. *in situ*, and the use of a motorised stage to allow the mapping of surfaces and the study of depth profiles of pigments on surfaces. The main difficulties arise with certain organic pigments which either fluoresce (or their supports or binders do), are photosensitive, or fail to yield a Raman spectrum owing to small particle size and/or high degree of dilution.

3.2.4.2 *Organic Coatings.*

Confocal Raman microscopy (Figure 3.7) is a powerful tool in the investigation of the depth homogeneity of curing of organic coatings. For example, Schrof *et al.* (2001) have successfully characterized UV-cured organic coatings by this means, helping to optimize UV formulations and curing conditions, e.g. UV dosage or temperature. This method combines chemical information with chemical imaging at 1 μm^3 resolution – well suited for a coating thickness of several tens of micrometres.

These authors detected the vibrational bands of reactive acrylate groups from a measuring volume of 1 μm^3, and calculated reaction conversions from reactive band ratios before and after curing (see Figure 3.9).

For standard coatings which can be penetrated by visible light, no sample preparation is required. For samples with high pigment content, microtomed sections are produced from which depth profiles may be obtained from a lateral scan. Schrof *et al.* (2001) employed a He–Ne laser (632.8 nm, 5 mW laser power at the sample surface) or a krypton ion laser, and selected a confocal volume of 1 μm^3 by a microscope objective with a magnification 100×, NA of 0.8 and a working distance of nearly 4 mm, using a confocal aperture of 75 μm. Computer-controlled movements of an *xy*-translational stage was employed to achieve mapping or depth-profiling.

Figure 3.10 is an example of depth profiles of curing conversion carried out with a polyesteracrylate formulation by these workers. While UV light is needed to initiate the curing process, it is known to be a source of long-term degradation of the cross-linked coating. The latter effect is reduced by the addition of light stabilizers, and UV absorbers (UVA) and sterically hindered amines (HALS) are mostly used for this purpose.

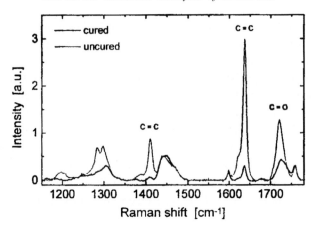

Figure 3.9. Raman spectra of urethane acrylates before and after UV exposure. After curing the reactive bands have partly vanished allowing quantitative determination of curing conversion. (After Schrof *et al.* 2001.)

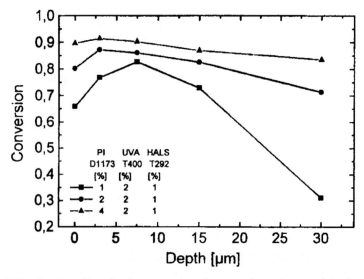

Figure 3.10. Depth profiles of curing conversion of an organic coating containing increasing concentrations of photoinitiator (PI), with fixed concentrations of light stabilizers (UVA and HALS). (From Schrof *et al.* 2001.)

Depth profiles of conversion for varying concentrations of photoinitiator (PI) are shown in the diagram, the concentrations of UVA and HALS have been kept constant. It is evident that lower PI concentrations are accompanied by oxygen inhibition at the coating surface and less bulk conversion.

3.3. LASER MICROPROBE MASS SPECTROMETRY

3.3.1 *Operating Principle*

The laser microprobe technique is based on performing a mass and intensity analysis of the ionic species formed when a pulse from a beam of laser light is focussed on to the surface of the specimen to form a probe ≥ 500 nm in diameter. Here it volatilizes a small region of the sample, releasing ions which are analysed in a time-of-flight mass spectrometer. The probe size is limited by the wavelength of the incident laser light, and so this analytical technique will therefore be at the upper end of our self-imposed scale of the degree of localisation of analysis.

All elements can be detected and measured with a sensitivity of the order of 1 ppm. The analysed volume is typically 0.1 cubic micrometers, so a sensitivity of 10^{-19} g is possible. Different isotopes can be distinguished, but the ionization probability is matrix dependent, which makes quantitative analysis difficult.

The technique is referred to by several acronyms including LAMMA (Laser Microprobe Mass Analysis), LIMA (Laser Ionisation Mass Analysis), and LIMS (Laser Ionisation Mass Spectrometry). It provides a sensitive elemental and/or molecular detection capability which can be used for materials such as semiconductor devices, integrated optical components, alloys, ceramic composites as well as biological materials. The unique microanalytical capabilities that the technique provides in comparison with SIMS, AES and EPMA are that it provides a rapid, sensitive, elemental survey microanalysis, that it is able to analyse electrically insulating materials and that it has the potential for providing molecular or chemical bonding information from the analytical volume.

3.3.2 *Experimental Details*

Odom and Schueler (1990) describe the basic components of the instrument, known as LIMA 2A or LAMMA 1000, depending upon the particular manufacturer. Figure 3.11 illustrates a schematic diagram of a reflection mode instrument.

The basic components include a Nd:YAG pulsed laser system which is coaxial with a He:Ne pilot laser and visible light optical system. The latter system enables the analytical area of interest to be located. The TOF-MS has a flight path of ~ 2 m in length, with an ion detection system that includes an electron multiplier detector, a multichannel transient recorder, together with a computer acquisition and data processing system.

The analytical area of interest is positioned in the focal spot of the He:Ne laser beam, the transient recorder is armed to record, and the Nd:YAG laser is fired. This laser pulse of between 5 and 15 ns duration produces a packet of ions that is accelerated from the sample surface and injected into the TOF-MS. All the

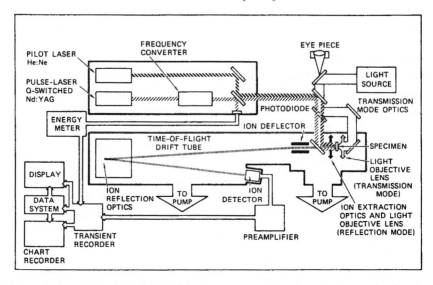

Figure 3.11. Schematic diagram of the model LIMA 2A laser microprobe mass spectrometer (Odom and Schueler 1990) to the number of ions detected.

accelerated ions have approximately the same kinetic energy, so ions of different mass will have different velocities. The time of arrival of the various ions at the detector is measured by the transient recorder, and the intensity of these digitized analogue voltages is proportional

In a *transmission mode* instrument, the Nd:YAG laser beam is focussed on the back side surface of thin samples ($\leq 1\,\mu m$ thick). A spot diameter of $\sim 0.5\,\mu m$ is possible, and commercial instruments of this configuration have been used primarily for biomedical and particle analysis applications.

3.3.3 Analysis of Single Airborne Particles by LIMS

Trimborn *et al.* (2000) have developed a mobile system for the on-line analysis of single airborne particles and for the characterisation of particle populations in aerosols, using a transportable laser mass spectrometer. A schematic diagram of their setup is shown in Figure 3.12.

The authors call their equipment 'LAMPAS 2' (Laser Mass Analyser for Particles in the Airborne State), the complete system being mounted on a movable rack on wheels with a height of 120 cm, a length of 170 cm and a depth of 80 cm. The total weight is about 250 kg. The operation is as follows:

On leaving the aerosol inlet (Figure 3.12) the particles pass a continuous laser beam and are detected by their light scattering signal. This signal is used to trigger a pulsed ionisation laser, and a single particle which is hit by the ionisation laser

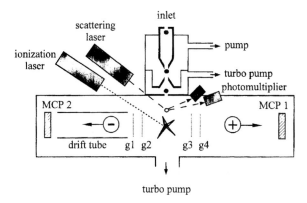

Figure 3.12. Schematic diagram of the instrumental setup of the mobile aerosol mass spectrometer 'LAMPAS 2'. (Trimborn *et al.* 2000.)

beam is partially vapourised and ionised. The positive and negative ions are analysed simultaneously by a bipolar TOF mass spectrometer containing two microchannel plate detectors (MCP1 and MCP2 in Figure 3.12). Spectra are recorded and stored by a personal computer.

The output pulse from the scattering laser triggers a delay generator which controls the delay time between the scattering pulse and the ionisation laser pulse. Particles with an appropriate velocity are hit by the ionisation laser pulse, and particle material is desorbed and ionised. For each acquired particle mass spectrum, the corresponding aerodynamic particle diameter is obtained from the chosen delay time and the known spacing between detection and ionisation laser beam (~ 1 mm). The software automatically switches from one size range to the next by changing the delay time for triggering the ionisation laser. Size analysis is thus performed by scanning across the chosen size ranges.

Trimborn *et al.* (2000) describe a four-week field campaign to characterise an aerosol in a particular area, where the LAMPAS 2 instrument was continuously analysing the size and composition of individual particles in five size ranges between 0.2 μm and 1.5 μm. Some 10,000 single particle spectra were recorded during the measuring period, and one example of these is shown in Figure 3.13.

The spectra are characterised by high signals of lithium, sodium, magnesium, aluminium, calcium and iron and their oxides. A high signal from hydroxyl ions was observed, which indicates a high water content of the particle.

3.3.4 Scanning Laser Microprobe Mass Spectrometry
Spengler and Hubert (2002) describe a confocal laser scanning microscope used in conjunction with a TOF mass spectrometer, and also possessing ion imaging

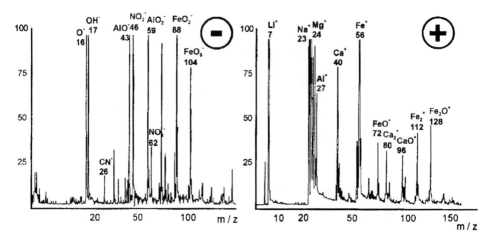

Figure 3.13. Examples of the positive ion and negative ion spectrum acquired from a single mineral dust particle (of aerodynamic particle diameter 1 μm). (Trimborn *et al.* 2000.)

capability. Yet another acronym is introduced by these authors to describe their instrument, namely SMALDI (Scanning Microprobe Matrix-Assisted Laser Desorption Ionisation) Mass Spectrometry. The set-up is shown in Figure 3.14.

The output of a Nd:YLF laser is focussed by a series of lenses to a spot size of 0.5 μm upon a sample which may be positioned by an *x-y-z* stepping motor stage and scanned by a computer-controlled high frequency *x-y-z* piezo stage. Ions are accelerated and transmitted through the central bore of the objective into a time-of-flight (TOF) mass spectrometer. The laser scans an area of 100×100 μm with a minimum step size of 0.25 μm. TOF mass spectra of each pixel are evaluated with respect to several ion signals and transformed into two-dimensional ion distribution plots.

The sample may be viewed by visible light (at a lower magnification) by using a diode laser for illumination and CCD camera. UV microscopic images may be obtained of exactly the same area as in the ion imaging operation by using a photo-multiplier tube (PMT) for light detection. The effective lateral resolution is in the range of 0.6–1.5 μm depending on sample properties, preparation methods and laser wavelength. Acquisition of an ion image with 1 μm resolution takes about 3–5 min, and a confocal optical image (Figure 3.7) with 0.25 μm resolution can be obtained in 20 s.

In non-highly focussed laser desorption ionisation, employing spot sizes in the range of 50–200 μm in diameter, the surface is deformed by an ablation volume of about 1 μm³ per pixel per laser pulse. But this ablated volume is spread over a large desorption area leading to ablation depths of the order of a few nanometres. In laser microprobing, the same ablation volume leads to ablation crater depths in the micrometer range.

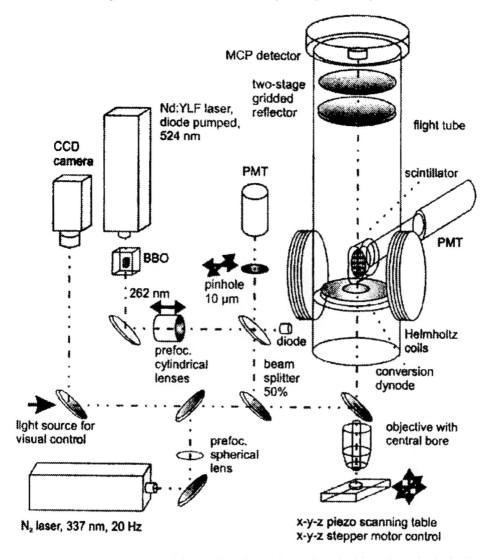

Figure 3.14. Schematic diagram of the scanning microprobe matrix-assisted laser desorption ionisation (SMALDI) mass spectrometer. (Spengler and Hubert 2002.)

Spengler and Hubert (2002) are concerned in their paper with biological applications of the SMALDI technique, and Figure 3.15 shows mass spectrometric images of a pine tree root. The sample was prepared by embedding the root in an epoxy resin, and a flat surface of the bulk sample was produced by cutting it with a glass knife. Ion signals of potassium and of calcium were imaged from the scanned sample in a cross-section of the root.

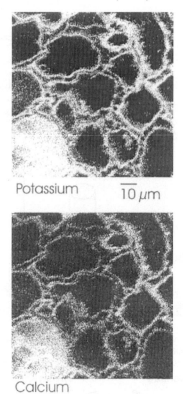

Figure 3.15. SMALDI image of a pine tree root, acquired at 262 nm wavelength. High ion intensities of potassium/calcium are coded in white, low intensities are coded in black. (Spengler and Hubert 2002.)

The size of the total image in Figure 3.15 is $100 \times 100\,\mu m$, and the images were scanned with lateral increments of $0.25\,\mu m$. The structures of the cell walls are clearly imaged, and show the expected high concentration of potassium and calcium in these areas.

REFERENCES

Afanasyeva, N.I. & Bruch, R.F. (1999) *Surf. Interf. Anal.*, **27**, 204.
Beitel, G.A., de Groot, C.P.M., Oosterbeek, H. & Wilson, J.H. (1997) *J. Phys. Chem. B*, **101**, 4035.
Berg, R.W. & Kerridge, D.H. (2000) *Proc. Molten State Chemistry*, **1**, 85.
Chabal, Y.J. (1988) *Surface Sci. Rep.*, **8**, 211.
Clark, R.J.H. (1999) *J.Molecular Structure*, **480–481**, 15.

Clark, R.J.H. & Gibbs, P.J. (1997) *Spectrochim. Acta*, **53A**, 2159.

Da Fonseca, C., Ozanam, F. & Chazalviel, J.-N. (1996) *Surface Science*, **365**, 1.

Eischens R.P. & Pliskin W.A. (1958) *Adv. Catal.*, **10**, 1.

Erné, B.H., Ozanam, F. & Chazlviel, J.-N. (1998) *Phys. Rev.Lett.*, **80**, 4337.

Ferraro, J.R. and Basilo, L.J. (1978, 1979) *Fourier Transform Infrared Spectroscopy: Application to Chemical Systems*, New York: Academic Press, Vols 1 and 2.

Loudon, R. (1964) *Adv. Phys.*, **13**, 423.

Loudon, R. (1965) *Adv. Phys.*, **14**, 621.

Schrof, W., Beck, E., Etzrodt, G., Hintze-Brüning, H., Meisenburg, U., Schwalm, R. & Warming, J. (2001) *Progress in Organic Coatings*, **43**, 1.

Spengler, B., and Hubert, M. (2002) *J. Am. Soc. Mass Spectrom* **13**, 735–748.

Stefanov, B.B., Gurevich, A.B., Weldon, M., Raghavachari, K. & Chabal, Y. (1998) *Phys. Rev. Lett.*, **81**, 3908.

Trimborn, A., Hinz, K.-P. & Spengler, B. (2000) *Aerosol Science and Technology*, **33**, 191.

FURTHER READING

Hirschmugl, C.J. (2001) *Surface Science*, **500**, 577.

McCreery, R.L. (2000) *Raman Spectroscopy For Chemical Analysis*, John Wiley & Sons, New York.

Odom, R.W. & Schueler, B. (1990) Laser Microprobe Mass Spectrometry: Ion and Neutral Analysis. In *Laser and Mass Spectrometry*, Ed. Lubman, D.M, Oxford University Press, Oxford, pp. 103–137.

Turrell, G. & Corset, J. (1996) *Raman Microscopy: Developments and Applications*, Academic Press, London.

Chapter 4
Ion Beam Probes for Surface Analysis

Chapter 4

Ion Beam Probes for Surface Analysis

4.1. INTRODUCTION

Ion Beam Analysis (IBA) utilizes high-energy ion beams to probe the elemental composition of the surface of a specimen in a non-destructive way. It can establish the composition as a function of depth to several microns, with a typical depth resolution of 10–20 nm. It is a fast and standardless technique which quantifies the absolute atomic ratios, and can also determine the film thickness.

IBA is a term that involves several specific techniques which we will consider in turn, namely:

Rutherford backscattering spectrometry (RBS) which analyses the elastic scattering of the particle beam from the target nuclei. Most RBS analyses use less than 2.2 MeV He^{++} beams.

Heavy ion backscattering spectrometry (HIBS) using ions heavier than He^{++} with accelerators and detectors similar to those for RBS.

Particle-induced X-ray emission (PIXE) in which electrons are ionised, and elemental specific X-rays are generated by the incident ion beam.

Particle-induced gamma-ray emission (PIGE) in which nuclei are excited and gamma-rays are generated by the ion beam.

Elastic recoil detection analysis (ERDA), or *Forward recoil spectrometry (FreS)* in which the samples are irradiated with high energetic heavy ions under grazing incidence conditions. *Hydrogen forward scattering (HFS)* is a special case of ERDA. In each of these situations, the energy transferred to the target nucleus is large enough to cause the *target nucleus* to recoil from the target surface. The nature of the emitted particle used as an analytical signal is thus obviously different from that of the incident one, and, unlike ion scattering techniques, it is an isotopically sensitive technique with an excellent mass resolution. These techniques use essentially the same apparatus as standard RBS.

Nuclear reaction analysis (NRA) also identifies emitted particles which are different from the incident ones. In order to avoid permanent radioactivity, the energy of the projectile is maintained below 6 MeV, so that it is used primarily to determine the concentration and depth of light elements ($Z < 9$) in the near surface of solids.

The technique is of high sensitivity (a few parts per million) and has a good depth resolution of a few nanometers, with a maximum depth of the few micrometres.

Charged particle activation analysis (CPAA) is based on charged particle induced nuclear reactions producing radionuclides that are identified and quantified by their characteristic decay radiation. CPAA allows trace element determination in the bulk of a solid sample as well characterization of a thin surface layer.

4.1.1 The Nuclear Microprobe

The first nuclear microbeam with a spatial resolution of 1 µm was built by Watt *et al.* (1981), and the first sub-micron instrument was built by Grime *et al.* (1987). Khodja *et al.* (2001) have published a description of the nuclear microprobe at the Pierre Süe Laboratory in France, which is a national facility dedicated to microbeam analysis. Its unique facility is that it is capable of analysing radioactive samples by means of a dedicated beamline. Figure 4.1 shows a schematic diagram of the apparatus.

Each of the two beamlines are focussed by a single electromagnetic quadrupole doublet, and a beam size of about $1 \, \mu m^2$ is achieved with a beam current of approximately 50 pA. Electrostatic deflectors are mounted to permit beam sweeping on the sample both to produce images and to limit heating effects.

The analysis chamber is such that PIXE, RBS, PIGE, NRA and ERDA are routinely performed simultaneously with the microprobe. A 4-axis micron-level goniometer permits precise positioning. The radioactive beamline enters a shielded

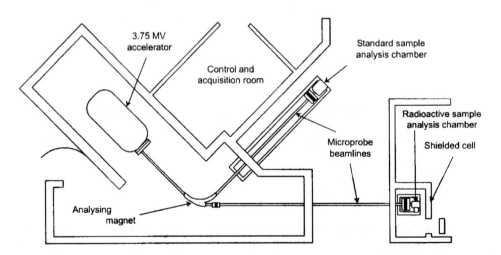

Figure 4.1. Schematic diagram of the Pierre Süe nuclear microprobe (Khodja *et al.* 2001).

cell, but the facilities available therein are broadly similar to those provided for standard samples.

The facility produces $^1H^+$, $^2H^+$, $^3He^+$ and $^4He^+$ beams that can be accelerated up to 3.6 MeV.

We will first consider, however, Secondary Ion Mass Spectroscopy (SIMS) in which both neutral and charged species are sputtered from the surface, and detected by means of a mass spectrometer. This involves ion beams of lower energy than in the techniques described previously.

4.2. SECONDARY ION MASS SPECTROSCOPY (SIMS)

The basic SIMS effect was discovered in 1910, although commercial development did not really take off until the 1960s. SIMS uses mass spectral analysis of surface atomic layers to show both elemental composition and molecular structure. The surface of the sample is subjected to bombardment by high energy primary ions (0.5–20 keV): this leads to the ejection (or *sputtering*) of both neutral and charged (+/−) species from the surface. The ejected species may include atoms, clusters of atoms and molecular fragments, and these are detected by means of a mass spectrometer. Analysis of the great majority of species which are not charged may also be accomplished by the technique of Secondary Neutral Mass Spectrometry (SNMS), which is experimentally more complex.

Figure 4.2 is a block diagram that illustrates the principle of the SIMS technique. The apparatus includes a primary ion source, a vacuum chamber where the objects under study are placed, a mass analyser and a secondary ion detector.

Wilson, Stevie, and Magee (1989) list the advantages and disadvantages of SIMS as follows:

Advantages

Detection limits of parts per million (ppm) atomic for most elements and parts per billion (ppb) atomic for favourable elements. 1 ppm is 0.0001%.

All elements detectable, even the 'light' elements from hydrogen to oxygen can be analysed and mapped.

Isotopes can be distinguished.

Depth resolution of 2–5 nm possible, and 10–20 nm typical.

Lateral resolution of 20 nm–1 μm depending on primary ion source.

Quantified using standards and relative sensitivity factors (RSFs).

Insulators analyzable.

Chemical information obtained from relative molecular ion abundances.

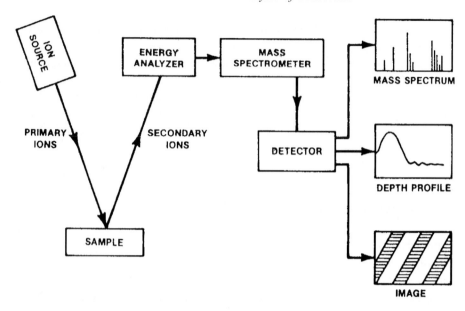

Figure 4.2.   Illustrating the SIMS technique.

Disadvantages

Mass interferences.
Secondary ion yields vary by more than six orders of magnitude over the periodic
   table.
Secondary ion yields are often matrix dependent.
Numerous secondary standards are required to quantify the data.
A flat surface required for best depth resolution and for ion microscopy
   Destructive analysis.

Microelectronics has been the field of highest utilization for SIMS because the
samples are usually flat, and the species of interest are often at such low concentration
that only SIMS has sufficient sensitivity to detect them. Other areas are metallurgy,
geology and biology. The technique is not limited to solid samples: the use of Xe^+ and
Cs^+ primary beams on liquids in a glycerol matrix has permitted the detection of
organic materials, particularly those species of high molecular weight, although this
application is not relevant to *local* chemical analysis, which we are considering here.
   Since ion beams (like electron beams) can be readily focussed and deflected on a
sample so that chemical composition imaging is possible. The sputtered particles
largely originate from the top one or two atom layers of a surface, so that SIMS is
a surface specific technique and it provides information on a depth scale comparable
with other surface spectroscopies.

SIMS is the most sensitive of all the commonly employed surface analytical techniques, because of the inherent sensitivity associated with mass spectrometry-based techniques, which can be of the order of parts per billion for some elements. There are several different variants:

Static SIMS, used for sub-monolayer elemental analysis. At the lowest current densities and hence the lowest rates of erosion, a monolayer on the surface has a lifetime of many hours. The surface is essentially unchanging during the experiment, but a vacuum system at a pressure of 10^{-10} mbar is needed to allow adequate time to complete the analysis. In favourable cases, as little as 0.1% of a monolayer of material can be detected.

Dynamic SIMS, used for obtaining compositional information as a function of depth below the surface. With higher etch rates, the surface is eroded much more quickly than in static SIMS, and depth profiling can be achieved since each successive layer of atoms can be analysed as it is 'peeled away'.

Imaging SIMS, used for spatially resolved elemental analysis. A focussed ion beam is rastered over the surface so that each point on the target is individually bombarded in turn, so that secondary ion emission is localized. The intensity of a particular secondary ion is monitored for each position of the primary beam and the result shown at the corresponding point of a synchronized oscilloscope display or computerized data system. In this way, pixel by pixel across the sample surface and in depth as the material is removed, three-dimensional information on the sample composition may be obtained.

We will now discuss the individual components of a SIMS apparatus, as illustrated in the block diagram of Figure 4.2.

4.2.1 The Primary Ion Source

An ideal ion source must possess high brightness, and must produce an ion beam of homogeneous composition with a small energy spread. The ion current density must be easily monitored and remain constant across the beam cross-section.

All ion guns feature a source of ions and some form of lens to extract the ions from the source chamber. More than one type of ion gun may be fitted to the instrument, as each will have strengths in certain areas of SIMS analysis.

4.2.1.1. Electron Bombardment Plasma Sources.
These gas-feed sources generally employ Ar or Xe at relatively low vapour pressures (10^{-3} mbar). A heated cathode is a common electron source, and these are accelerated towards an anode to give them

the required energy to form a plasma by ionizing interaction occurring during their flight to the anode. Ions are extracted from the plasma through an exit aperture to give an ion beam with an energy spread of < 5 eV.

Inert gas beams allow the chemistry of a surface to be studied by SIMS without modification by the bombarding species. The achievable values of source brightness allow μA currents into spot diameters of approximately 50 μm for dynamic SIMS, or nA currents into spot diameters < 5 μm for imaging SIMS. For greater spatial resolution a different, higher-brightness source must be used.

4.2.1.2. *Liquid Metal Sources.*

The source feed is a metal of low melting point – Ga and In are commonly employed. It is introduced as a liquid film flowing over a needle towards the tip whose radius is relatively blunt (10 μm). The electrostatic and surface tension forces form the liquid into a sharp point known as the Taylor cone. Here the high electric field is sufficient to allow an electron to tunnel from the atom to the surface, leaving the atom ionized.

These sources have high brightness and can deliver currents of over 100 μA, although the energy spread is quite high (10 eV) at high beam currents. Currents of the order of nanoamperes may be focussed into spot sizes from 0.1 to 0.02 μm, and such submicron imaging is only achievable using such sources. These beam currents are too small to allow depth profiling of large areas (100 μm × 100 μm), though small area dynamic SIMS is possible.

4.2.1.3. *Surface Ionization Sources.*

In this system, a low ionization potential atom (e.g. caesium) is adsorbed on a high work function metal (e.g. iridium). The temperature is raised so that the rate of desorption exceeds the rate of arrival of the atoms at the surface, and the Cs is then desorbed as ions with very small energy spread (< 1 eV). The spot size – current characteristics of these sources lie between liquid metal and plasma discharge sources.

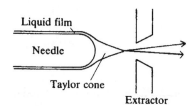

Figure 4.3.   Diagram of the extraction region of a liquid metal ion source.

4.2.1.4. Ion Gun Components. All ion guns feature some form of lens to extract the ions from the source chamber. A potential difference is applied between the extractor and sample (0.5–30 kV) which determines the final energy of the ions and hence the beam energy. A second lens is usually incorporated to give a degree of focussing at the target, and *xy* deflection plates permit positioning of the beam or raster scanning.

A mass-energy filter may be included to remove contaminant ions, and multiply charged and clustered ions of the main beam species. Ions of the required mass and energy pass through the filter, while unwanted species are deflected out of the beam. Prior to final focussing, the beam is deflected through a few degrees, and any neutral particles will be undeviated and are therefore separated out.

4.2.2 The Mass Analysing System

The bombarding of the specimen surface by the primary beam of high energy ions leads to the ejection (sputtering) of both neutral and charged (+/−) species from the surface. The ejected species may include atoms, clusters of atoms and molecular fragments. The ions enter an extraction lens and the polarity of the applied voltages determines the polarity of the secondary ions that enter the analyser.

The analyser will always be preceded by some form of collection optics, and followed by an ion detector (usually a channel electron multiplier which converts ions into electron showers). There are three types of analyser for use in SIMS spectrometers, the *magnetic sector* instrument, the *quadrupole* analyser and *time-of-flight (TOF)* systems.

4.2.2.1. The Magnetic Sector Analyser.

This separates the beam of secondary ions into its component parts according to their mass/charge ratios. The arrangement is illustrated diagrammatically in Figure 4.4, which represents a longitudinal cross-section of the apparatus, showing the mass dispersion. The magnitude of the force

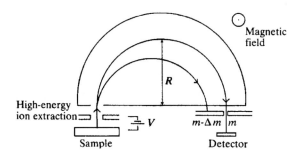

Figure 4.4. Illustrating the operation of a magnetic sector mass analyser.

experienced by the charged particle as it moves through the magnetic field depends on the velocity of the ion. The force acts orthogonally to both the direction of motion and the magnetic field, resulting in a circular trajectory.

The field strength is scanned by an electromagnet, and the dispersion of adjacent masses (i.e. the resolution) decreases with increasing ion mass. The high secondary ion extraction voltage employed results in efficient transmission of secondary ions from the sample surface to the detector, although it is difficult to analyse samples with surfaces that are fractured or rough.

Magnetic sectors have medium transmission (10–50 percent), large mass range ($m/z > 10000$, where m is the mass of the ion and z the charge) and excellent mass resolution ($m/\Delta m > 10^4$). This high mass resolution makes these analysers appropriate for SIMS analysis in semiconductor research.

4.2.2.2. The Quadrupole Mass Analyser.

The radio frequency (RF) quadrupole mass analyser is more compact than the magnetic sector analyser, and easy to use. Figure 4.5 illustrates the principle of its operation. It contains four rods connected together as two opposite pairs. A potential consisting of a constant DC component plus an oscillating RF component is applied to one pair of rods, and an equal but opposite voltage is applied to the other pair. A rapid periodic switching of the field sends most ions into unstable oscillations until they strike the rods and are not transmitted. Ions with a certain mass/charge ratio follow a stable trajectory and are transmitted. By increasing the DC and RF fields at a constant ratio between them, this stable condition is satisfied for ions of each ascending mass in turn, allowing the collection of a complete mass spectrum.

Because these analysers do not employ magnets, peak switching for selected ion monitoring can be done more quickly without hysteresis effects, which makes this system ideal for depth profiling, where it is necessary constantly to switch among masses. These instruments do have the disadvantage of loss of transmission and mass

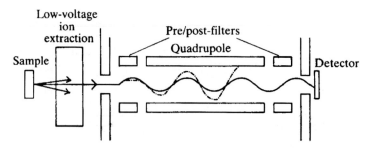

Figure 4.5.   Longitudinal section through a quadrupole mass filter (showing two of the four rods), illustrating stable and unstable ion trajectories.

resolution with increasing mass, and they do not achieve high mass resolution (10^2–10^3). The mass range (m/z) is < 1000, and the ions for analysis must be of low energy ($\sim 10\,eV$) to allow an adequate length of time to be spent inside the filter ($\sim 150\,mm$ in length) for effective separation.

4.2.2.3. The Time-of-Flight Analyser.

TOF analysers are used with secondary beams that are generated in short pulses (< 10 ns in duration). The ions are accelerated into a drift tube (Figure 4.6), and if all the ions have the same energy, the ions become separated according to the time taken to traverse the drift space. Heavier ions take longer to reach the detector than do lighter ions. A time-sensitive detection system will then generate a mass spectrum.

The secondary ions are accelerated into the analyser to about 5 kV so that the flight time over a distance of $\sim 2\,m$ is reasonable ($\sim 100\,\mu s$). This time, plus a similar time for data processing, must elapse before the next pulse of primary ions is sent. The arrival time of the ions at the detector has to be measured to an accuracy of at least 1 ns. TOF analysers generally have a high transmission (50–100 percent) and a mass range limited to between $m/z = 5000$ and 10,000 by practical considerations. A high mass resolution is also achievable with this type of analyser, being at least 10^3. They have been principally used to obtain mass spectra and have had limited application for depth profiling, which would be time consuming.

4.2.3 Chemical Analysis

The principal advantages of SIMS, in both its static and dynamic forms, are its surface sensitivity and its very low detection limits for impurities. Only a very small proportion of the detected ions come from the second or lower layers of the materials being analysed. With regard to quantitative analysis, Vickerman (in Vickerman,

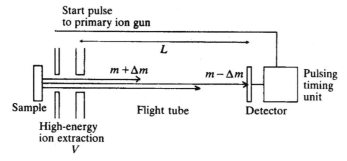

Figure 4.6. Illustrating the principle of the TOF mass analyser, showing a longitudinal cross-section of the flight tube and demonstrating mass dispersion.

Brown and Reed (1989) has reviewed the theoretical models of SIMS and he concludes that the lack of a fully comprehensive theory has not inhibited the successful empirical application of SIMS to many problems.

There are two parameters which are of primary importance in the analysis of SIMS data – the relationship between the secondary ion current and the elemental concentration in the sample surface, and the depth scale.

4.2.3.1. Secondary Ion Yields. The most successful calculations of secondary in yields are based on the local thermal equilibrium (LTE) model of Andersen and Hinthorne (1973), which assumes that a plasma in thermodynamic equilibrium is generated locally in the solid by ion bombardment. Assuming equilibrium, the law of mass action can be applied to find the ratio of ions, neutrals and electrons, and the Saha–Eggert equation is derived :

$$n_{\mathrm{e}}n_{\mathrm{x}}^{+}/n_{\mathrm{x}}= \{2Z^{+}(T)/Z^{0}(T)\}[2\pi mkT]^{3/2}/h^{3}]\exp[(E_{i} - \Delta E_{i})/kT]$$

where n_{e}, n_{x}^{+} and n_{x} are the absolute electron, ion and neutral atom densities respectively of element x in the surface plasma, Z^{+}, Z^{0} and 2 are the partition functions of ions, neutrals and electrons respectively and are temperature independent, E_{i} is the ionization energy, ΔE_{i} is the depression of ionization energy in the plasma and T the plasma temperature, m is the electron mass, h is Planck's constant and k is Boltzmann's constant.

Using the temperature and electron density as fitting parameters, within a range established from measurements from known samples, the ratio $n_{\mathrm{x}}^{+}/n_{\mathrm{x}}$ may be obtained and the concentration of element x calculated using

$$X^{\mathrm{x}}= n_{\mathrm{x}}^{+}(1 + n_{\mathrm{x}}^{+}/n_{\mathrm{x}})$$

The values of T and n_{e} are iterated until a suitable convergence limit is achieved. Werner (1980) has reviewed other quantification models, some leading to a more satisfactory one-parameter equation.

If homogeneous samples of a similar composition are to be analysed on a routine basis, a data bank of *sensitivity factors* can be established for the elements and host matrices being studied. The ion current recorded for element A emitted from matrix M is written as

$$I^{\mathrm{A}}= [\mathrm{d}I_{\mathrm{M}}^{\mathrm{A}}/\mathrm{d}X^{\mathrm{A}}]X^{\mathrm{A}}$$

Where X^A is the concentration of element A and dI_M^A/dX^A is the *absolute sensitivity factor* of element A in matrix M. A calibration curve is established using known samples for each element in each matrix. Unknown concentrations may then be determined provided all instrumental factors and experimental conditions are held constant, which conditions may be difficult to fulfil.

Relative sensitivity factors (RSFs) may be more readily applied if the intensities are not referred to an internal reference standard, when the concentration is given by

$$X_M^A = S_M^A I^A / f^A I_M$$

where S_M^A is the relative sensitivity factor for element A in matrix M, I^A and I^M are the measured secondary ion currents of elements A and M respectively, and f^A is the isotopic abundance of element A. There are published relative sensitivity factors available for SIMS, and the method can give accuracies of around 10%.

The *indexed relative sensitivity factor* approach obviates the necessity of measuring the relative sensitivity factors from all possible matrices, by transferring relative sensitivity factors for elements between different matrices by using the matrix-dependence of characteristic intensity ratios in the spectra. Calibration curves are constructed relating RSFs for an element in a matrix to the matrix ion species ratio (e.g. M^{2+}/M^+ for element M) generated from a single standard.

4.2.3.2. Sputter-Ion Depth Profiling.

Although it is essentially a destructive technique, SIMS depth profiling is rapid, and possesses parts per million or even parts per billion sensitivity to *all* elements and isotopes, coupled with a depth resolution of a few nanometres. Concentration–depth plots can be accurate to 5%.

The depth resolution (i.e. the ability to discriminate between atoms in adjacent thin layers) is limited by the primary beam causing redistribution of target atoms prior to their emission as ions, and to segregation and radiation-enhanced diffusion processes. The local topography can also lead to a loss of depth resolution with sputter depth.

As a first approximation, a simple linear relation may be assumed to convert the time of sputtering to depth. Thus for a constant sputtering rate, the depth of erosion z will be given by:

$$z = (MSJ_p t)/(\rho Ne)$$

where M is the mass1 number of the sputtered species, S is the sputtering yield in atoms per incident ion, J_p is the primary ion current density, t is the time of sputtering, ρ the density, N is Avogadro's number and e the electronic charge.

The value of S will be a complex function of the sample composition, and of the incident ions and their energy, mass and angle of incidence. If the crater depth can be measured independently, the above equation can be used to obtain the value of S.

4.2.4   Examples

4.2.4.1. Imaging SIMS.   Steeds *et al.* (1999) included this technique in their study of the distribution of boron introduced into diamond, where it is a well-established dopant that controls the electrical conductivity. SIMS was performed with a field-emission liquid gallium ion source interfaced to a magnetic sector mass spectrometer capable of about 0.1 μm spatial resolution.

Diamond was doped by chemical vapour deposition to levels of 10^{19}–10^{21}/cm^3. A bias was applied to achieve efficient extraction of the secondary B ions, and these were detected to map out the variation of B concentration from one area to another, as illustrated in Figure 4.7.

4.2.4.2. Dynamic SIMS.   SIMS depth profiling is normally used to determine the concentrations in the range 10^{13}–10^{20} atoms/cm^3 lying at depths of up to 10 μm.

Accurate chemical analysis of semiconducting samples has been a main application of the technique. It has a sensitivity of parts per billion of dopant in favourable cases, with a depth resolution usually between a few nm and a few tens of nanometres. A high primary beam current will favour a high sensitivity, but the destructive nature of the method means that these conditions will not lead to the best depth resolution. The speed of depth analysis is usually between 0.1 and 10 μm h^{-1}, but again a high sputter rate may give an unacceptably large depth increment between data points.

McPhail (1989) gives a detailed account of the experimental approach to depth profiling of semiconductors, including the quantification of the data. He illustrates the analysis of a silicon epilayer grown by molecular beam epitaxy (MBE) in which 11 boron-rich layers were incorporated by co-evaporation of boron. The intended structure is shown in Figure 4.8, and it was desirable to establish the concentration of boron in the layers, the inter-peak concentration and the sharpness of the doping transitions.

The sample was analysed with a primary ion beam of $^{16}O_2^+$ extracted from an oxygen cold cathode discharge type source. A well-focussed 200 nA spot was attained at 4 keV per molecular ion. The beam was scanned over a square of side 400 μm to produce a uniform primary beam current density and thus a flat-bottomed crater. In order to eliminate crater-edge effects, the counting system was only enabled when the centre of deflection of the beam was in a central area 125 × 125 μm.

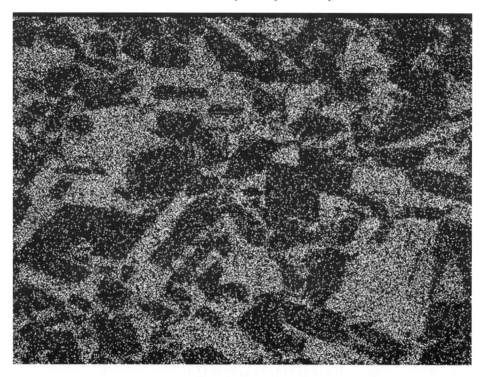

Figure 4.7. Boron scanned ion beam mass spectrometry (SIMS) image of a B-doped diamond specimen. The image is ~ 430 μm in width, and the variation of B intensity from one grain to another is approximately a factor of 8. (Reproduced by permission of Professor J. W. Steeds.)

An ion-implanted standard and the MBE sample were depth profiled under the same conditions, and the secondary ions were analysed in a quadrupole mass spectrometer. The data from the ion-implanted standard was used to find the useful ion yield and thus the instrumental sensitivity for boron-in-silicon in the MBE sample. The quantified data appear in Figure 4.9.

A comparison of Figures 4.8 and 4.9 illustrates the power of this analytical technique. McPhail comments that the measured dynamic range is poorer than might have been expected and it appears to decrease with depth, and the causes of this are discussed.

4.2.4.3. SSIMS. The study of surfaces of catalytic interest was one of the earliest applications of static SIMS, and Vickerman *et al.* (2000) have recently reviewed recent applications of the technique in this area. As an example of this

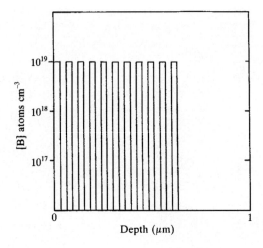

Figure 4.8.   Intended doping distribution for a boron-in-silicon test structure. (McPhail 1989.)

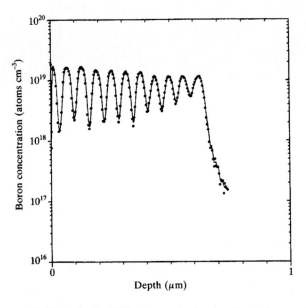

Figure 4.9.   The quantified data for the MBE test sample, showing the total concentration of boron (both isotopes) as a function of depth (McPhail 1989.)

approach, the studies by these authors of auto-exhaust catalysts will be outlined below.

These catalysts are able to oxidize simultaneously the unburnt hydrocarbons and carbon monoxide to H_2O and CO_2, and to reduce NO_2 emissions to N_2. They are composed of a ceramic monolithic support that is covered in a washcoat of γ-alumina and ceria which is also impregnated with < 0.2 at.% of platinum and rhodium.

SSIMS has been applied to study of the chemical poisoning and thermal deactiviation of the catalyst that takes place with increasing distance travelled by the car. Figure 4.10 shows the positive and negative SIMS spectra obtained from a fresh fully formulated auto-exhaust system.

The spectrum was obtained by time-of-flight SSIMS from a catalyst consisting of 0.2% Pt/Rh(5 : 1) + $CeO_2/\gamma-Al_2O_3$ on a cordierite monolith. The authors suggest that the absence of Ce_2O_x or Ce_3O_x species (which were observed in a model catalyst composed of a Ce_2O – only washcoat) may reflect the high dispersion of ceria in alumina and may suggest that the ceria is in a slightly reduced state. The absence of any Pt species in the spectrum is a notable feature, but it was found that the use of a Cs^+ primary ion source enhanced the yield of Pt species as negative ions (see Figure 4.11).

Figure 4.11 reveals that Pt is present on the surface of the catalyst as an oxide, in combination with hydrocarbon species (a contaminant during sample preparation) and as a chloride (derived from the Pt precursor, chloroplatinic acid). The results show the composition of the washcoat to be Pt and Rh on alumina and ceria.

Table 4.1 lists the ions observed from a 121,000 km vehicle-aged catalyst, and the poisons to which they can be attributed.

4.3. RUTHERFORD BACK-SCATTERING SPECTROMETRY (RBS)

RBS is based on collisions between atomic nuclei, and it involves measuring the number and energy of ions in a beam which backscatter after colliding with atoms in the near-surface region of a sample. The use of scattering as an analysis tool led to the first *in situ* chemical analysis of the lunar surface during the landing of Surveyor V. The use of particle accelerators as an α-source was the next powerful step made in Chalk River (Canada) and Årus (Denmark).

The RBS method is represented schematically in Figure 4.12.

Typically $^4He^{++}$ in the energy range 1–3 MeV is employed as the incident beam, and ions are detected with a solid state detector of energy resolution around 20 keV after scattering through a relatively large angle of 160°–170°. The detector counts the number of scattered particles and measures their energy, and the information obtained can be interpreted to give data on the composition of the sample, the distribution of components within it, as well as the sample thickness. The major development of RBS as an analytical tool was done at IBM (Ziegler)

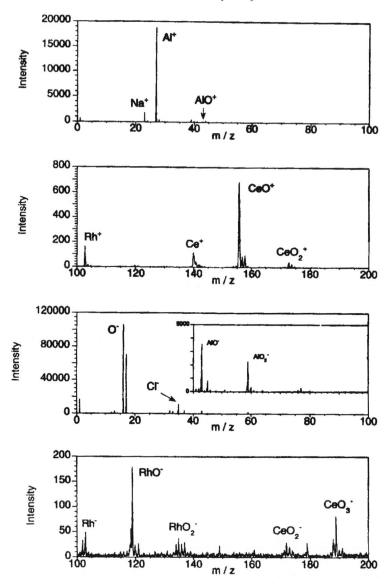

Figure 4.10. SIMS spectra obtained from a fresh auto-exhaust catalyst.
(Courtesy Vickerman *et al.* 2000.)

and Cornell (Mayer). Grant (1989) gives an introduction to the physical concepts of the technique, and a more comprehensive coverage is provided by Chu *et al.* (1978).

It is possible to measure nearly any type of sample for almost any element with little or no preparation, provided that the sample is stable in a vacuum. RBS is not

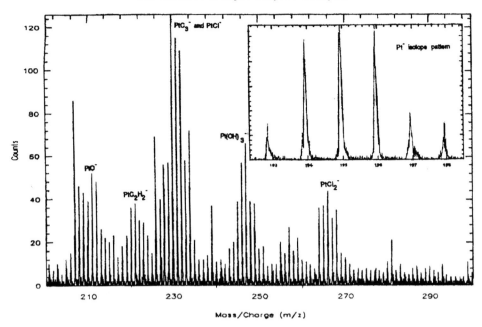

Figure 4.11.   Cesium ion initiated static SIMS spectra of platinum species on fresh auto-exhaust catalyst. (After Oakes *et al.* 1996).

Table 4.1. The principal secondary ions observed from a 121,000 km vehicle-aged catalyst

Positive			Negative		
m/z	Ion	Possible origin	*m/z*	Ion	Possible origin
23	Na^+	Crude oil	31	P^-	Oil additive
24	Mg^+	Ceramic	47	PO^-	Oil additive
28	Si^+	Ceramic	63	PO_2^-	Oil additive

good for unknown mixtures of heavy elements, because the mass resolution drops rapidly at high masses. Besides solid state samples, examples include powders, air particulates or chemical precipitates on filters, and biological samples such as bone, teeth and shells. Generally 10 mg of sample are required, and a maximum sample size of 50×100 mm is possible.

The method is essentially non-destructive, although in some circumstances a target may suffer radiation damage. For example the lattice site position of some dopants in semiconductors can be influenced by RBS analysis. With a typical analysis time of less than 30 min, it is a relatively quick method. Its most important characteristic is

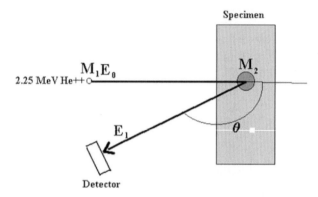

Figure 4.12. Schematic diagram of the RBS method.

that RBS can provide quantitative measurement of film thickness and impurities without the need for calibrated standards.

4.3.1 Instrumentation

The three main components of an RBS instrument are a source of helium ions, an accelerator and a detector to measure the energies of the backscattered ions.

Many RBS installations use a *tandem accelerator*, producing a 2.25 MeV He^{++} beam by removing three electrons from He^-.

4.3.1.1. Source of He^-.

The source operates in two stages. Firstly He^+ or He^{++} comes from a plasma ion source by introducing He into a low voltage arc. The helium plasma is geometrically and magnetically confined, and ions are extracted by a strong electric field. In a second stage, the He^+ beam is passed through a hot alkali metal (e.g. rubidium) vapour where a charge exchange occurs, since the alkali metal has sufficient reducing power to form He^-. The efficiency of the charge exchange process is such that 1 mA of He^+ leads to 1 μA of He^-. The negative ions are extracted and injected into the tandem accelerator at 20 to 30 keV.

4.3.1.2. The Tandem Accelerator.

As indicated in the diagram of Figure 4.13, a tandem accelerator uses a positive terminal located in the centre of the device. Negatively charged He^- particles are injected into the accelerator and attracted to the terminal, where a 'stripper element' removes two or more electrons from each

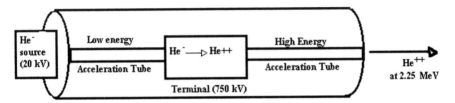

Figure 4.13. Schematic diagram of a tandem accelerator source of high energy alpha particles.

particle. The positive terminal now repels the resulting positive ion, so the particle acquires energy both before and after the terminal. An RBS installation might use an accelerator of this type to produce a 2.25 MeV He^{++} beam by removing three electrons from He^{-} at the positive 750 kV terminal. In fact, at the present time, more laboratories use single-ended accelerators than tandem ones.

4.3.1.3. Focusing Elements. Two ion optical components are usually placed between the accelerator and the sample chamber. A magnetic field separates any He^{-}, He, or He^{+} from the He^{++} beam. A quadrupole magnetic lens then shapes the beam and focusses it into the sample chamber.

4.3.1.4. The Scanning Microprobe. Some research establishments provide a scanning proton microprobe system, with a beam spot diameter typically ~0.01–0.02 mm. Over the past 20 years the minimum probe size has stayed around 1 µm with only a few groups reporting a sub-micron probe size around 0.5 µm in the quest for improved spatial resolution. This finely focussed beam can, when stationary, measure concentrations of elements in a sample, or it can scan across a line, or over a two-dimensional area of up to 5×5 mm to study elemental variations over a region. It can be programmed to analyse for up to 20 elements simultaneously, and it seldom damages or destroys the sample it is analysing.

4.3.1.5. Sample Chamber and Detector. The pressure in the sample chamber is typically 10^{-6} torr, although UHV may be required for some experiments. The samples are usually mounted on a five-axis goniometer, so that a series of samples may be loaded and analysed sequentially. The goniometer can tilt and rotate the samples relative to the direction of the incident beam. Comparing spectra obtained at different incident and exit beam angles provides fuller characterization of the sample composition as a function of depth. The samples can be electrical insulators (in contrast to most charged particle analytical methods), since the high energy (rigid) ion beam is virtually unaffected by sample charging. If the sample were to charge to a few kilovolts, only slight perturbations of the spectra would result, because the charging effect is but a small fraction of the megavolt energy of the incident ion beam.

Surface barrier silicon detectors are used in RBS. Two such detectors (each of area $\sim 25 \, mm^2$) are commonly employed at a distance $\sim 150 \, mm$ from the target–one at a scattering angle of $\sim 170°$, i.e. almost total backscattering, and one at grazing exit ($\sim 9°$), which may be used for near-surface analysis. The angular spread of the beam at the detector can be reduced by placing slits in front of the detector. The high-energy charged particles entering the detector produce electron–hole pairs in these semiconductor diode detectors which are operated with an electrical potential of 10–100 V between the front and back surfaces. The electron–hole pairs produce a pulse of charge proportional to the energy of the charged particle. Particle arrival times at the detector occur at random times, leading to the possibility of two pulses being sufficiently close together to be treated as a single pulse (peak pile-up). Detector dead time is the minimum time between successive ion arrivals if they are to be measured separately. Peak pile-up limits the rate at which RRS data collection can occur.

The signals from the detectors are amplified to create a voltage pulse with amplitude proportional to the energy of the charged particle. Data acquisition, storage and display is effected by an MCA providing pulse-height analysis.

4.3.2   RBS Theory

4.3.2.1. The Kinematic Energy Loss.   The elastic collisions between two particles can be solved by applying the principles of energy and momentum conservation. For a simple collision, the ratio (K) of ion energy E_0 before collision and E_1 after collision is a function of the masses of the projectile (M_1) and target (M_2) and the scattering angle θ, (Figure 2.1) therefore:

$$E_1 = KE_0 = \left[\frac{(M_2^2 - M_1^2 \sin^2\theta)^{1/2} + M_1\cos\theta}{M_1 + M_2} \right]^2 . E_0$$

$$\approx \left(\frac{M_2 - M_1}{M_2 + M_1} \right)^2 . E_0 \quad \text{for } \theta \approx \pi$$

K denotes the kinematic factor for the elastic scattering process. Thus, in RBS, a projectile of known mass and known energy is employed, and by measuring the energy of particles scattered at an angle θ the unknown mass of the target can be found.

If a target contains two masses that differ by a small amount ΔM_2 the difference in the energy of the projectile after collision is given by

$$\Delta E_1 = E_0(\mathrm{d}K/\mathrm{d}M_2)\Delta M_2.$$

The largest change in K for a given ΔM_2 occurs when $\theta = 180°$, and so large scattering angles maximise the mass discrimination of the technique.

If ΔE_1 is set equal to δE, the minimum energy separation that can be resolved experimentally, then the mass resolution of the system, δM_2, is:

$$\delta M_2 = \frac{\delta E}{E_0(\mathrm{d}K/\mathrm{d}M_2)}$$

The mass resolution at the sample surface is usually determined primarily by the detector resolution.

4.3.2.2. Scattering Cross Sections.

Particles are scattered into the solid state detector that subtends a small solid angle Ω (typically less than 10^{-2} sr). The number of counts, H, registered by the detector, and thus the height of the spectrum, is given by

$$Y = Q.Nt.(\mathrm{d}\sigma/\mathrm{d}\Omega).\Omega$$

where Q is the number of particles that strike the target, N the atomic concentration of target atoms, t the thickness of the layer probed by the beam (so Nt is the areal density of atoms) and $\mathrm{d}\sigma/\mathrm{d}\Omega$ is the average differential scattering cross-section. For small values of Ω this is usually referred to as the *scattering cross-section*, denoted by σ, so the yield can be written:

$$H = Q.Nt.\sigma.\Omega$$

and it is this which governs the height of the spectrum detected for a particular energy range.

If the force between the beam particle and the target nucleus is assumed to be the Coulomb force, the basic equation for the differential scattering cross-section is given by Rutherford's formula:

$$\frac{\mathrm{d}\sigma}{\mathrm{d}\Omega} = \left(\frac{Z_1 Z_2 e^2}{2E \sin \theta}\right)^2 \frac{\left\{\cos \theta + [1 - (M_1/M_2)^2 \sin^2 \theta]^{1/2}\right\}^2}{[1 - (M_1/M_2)^2 \sin^2 \theta]^{1/2}}$$

with

$$\sigma = (1/\Omega) \int_\Omega (d\sigma/d\Omega)\, d\Omega$$

Z_1 and Z_2 are the atomic numbers of the projectile and target atom respectively. The scattering cross-section is basically proportional to the square of the atomic number of the target atom, so RBS is over 100 times more sensitive for heavy elements than for light elements, due to the larger scattering cross-sections of the heavier elements. The proportionality to $1/E^2$ indicates that the scattering yield rises with decreasing bombarding energy.

The scattering cross-section is modified strongly by nuclear effects for energies approaching the Coulomb barrier (i.e. high energy ions hitting low mass nuclei). This is a major effect for proton RBS, and can also be significant for light element RBS with alpha-particles.

4.3.2.3. Stopping Cross-Section.

When the probing particles penetrate the target, the projectile energy dissipates due to interactions with electrons (*electronic stopping*) and due to glancing collisions with the nuclei of the target atoms (*nuclear stopping*). Thus a particle which backscatters from an element at some depth in a sample will have less energy than a particle which backscatters from the same element on the sample surface. The amount of energy lost by probing particle per distance travelled depends on the particle, its velocity, the elements in the sample and the density. This energy loss enables layer thicknesses to be measured by RBS, in a process known as *depth profiling*.

The majority of energy loss is caused by electronic stopping: nuclear stopping only contributes significant energy losses at low particle energies. The ratio of energy loss to two-dimensional atom density for a given material is known as its *stopping cross-section* (ϵ), commonly measured in units of eV-cm. Theoretical predictions of stopping power are complicated, and empirical stopping powers are often used in RBS calculations.

When a beam slows down in a target composed of more than one element, the energy loss can be calculated using Bragg's rule , which states that the total energy loss ϵ^{AB} in a compound $A_m B_n$ is given by

$$\epsilon^{AB} = m\epsilon^{A} + n\epsilon^{B}$$

where ϵ^A and ϵ^B are the stopping cross-sections of the atomic constituents A and B.

The stopping power of a material is defined as the energy loss per distance travelled in the material, dE/dz, and the Bragg rule may be expressed more generally, if the stopping cross-sections for each element are known:

$$\frac{dE}{dz} = N \sum c_i \epsilon_i$$

where N is the atomic target density, c_i and ϵ_i are concentration and stopping cross-section for element i, respectively. The Bragg rule assumes that the interactions between the incident ion and target atoms do not depend on the local environment. The different bonding states which can be present in compounds often cause deleterious effects for other analytical methods such as SIMS or AES, but due to the high energy of the probing particles in RBS compared to the very small binding energies, this does not present a problem, which is one reason that quantitative analyses can usually be performed by RBS without the use of standards.

4.3.2.4. Straggling. The essence of RBS is to measure the energy of the scattered beam and to calculate thereby the depth and/or mass from which scattering occurs. Any uncertainty in particle energy leads to a reduction in the precision with which mass and depth analysis can be achieved.

The process whereby a probing particle loses energy involves a large number of interactions with individual atoms along its trajectory. This causes statistical fluctuations in the energy loss process which limit the energy resolution which can be achieved for atoms backscattered from larger sample depths, and this phenomenon is called *energy straggling*.

The Bohr (1953) expression for energy straggling is:

$$\Delta E^2_{straggling} = 4\pi Z_1^2 Z_2 e^4 z$$

and its magnitude is dependent only on the atomic numbers and increases with the penetration depth, z.

Straggling thus limits the depth and mass resolution for features buried within the target material. The depth resolution, Δz depends on the stopping power, dE/dz, the detector resolution, ΔE_{det} and the beam energy spread, ΔE_{beam} of the incident particles:

$$\Delta z \frac{dE}{dz} = \sqrt{\Delta E^2_{beam} + \Delta E^2_{straggling} + \Delta E^2_{det}}$$

Usually the beam spread can be neglected, since it is much smaller than the energy resolution of the detectors.

4.3.2.5. Channelling.   When the incoming beam is aligned with any low-index axis or plane in a crystal it can be channelled, so that the probe atoms are steered down the channels. Under these conditions the backscattering yield will decrease to a few percent of its original non-oriented value (the 'random yield').

   Channelling effects can provide two types of information in RBS experiments. If a detector is adjusted to have an energy window corresponding to a chosen atomic species, a specimen tilt-through over a channelled direction brings information on the perfection of crystallinity of the target and also on the lattice location of dopants or impurities. The yields vs. tilt-through curve has a minimum in the channelled direction, and the smaller this minimum yield, the more perfect is the crystal.

   With regard to impurity atoms, the tilt-through experiment for a substitutional impurity will again show a minimum in the channelling direction, except that the energy value will be different. An altered FWHM value for the line in the spectrum in this case gives information about the exact lattice position and possibly the size of the impurity. An interstitial impurity will however be visible for any angle of incidence of the beam, and the degree of RBS will not be influence by the angle of incidence, so the tilt-through experiment gives information on the substitutionality/interstitial property of the dopants.

4.3.2.6. Layer Thickness Measurements.   The energy E_1 of a particle can be related to the depth z at which scattering occurs, as indicated in Figure 4.14. A penetrating particle with a path length $z/\cos\theta_1$ loses energy ΔE_{in}, and suffers a further energy loss ΔE_{out} as it travels a distance $z/\cos\theta_2$ to leave the target. The final energy of the emerging particle is therefore

$$E_1 = K(E_0 - \Delta E_{in}) - \Delta E_{out}$$

The energy difference ΔE between particles scattered at the surface and particle scattered at a depth z is therefore

$$\Delta E = K.\Delta E_{in} + \Delta E_{out} = z\left\{\frac{K}{\cos\theta_1} \cdot \frac{dE}{dz}\bigg|_{E_0} + \frac{1}{\cos\theta_2}\frac{dE}{dz}\bigg|_{E_0}\right\}$$

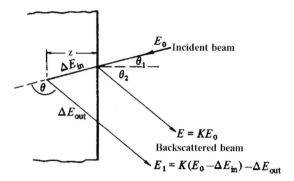

Figure 4.14. Showing the trajectories and energies of particles backscattered at the surface and at a depth z perpendicularly below the surface.

Any energy difference can thus be converted into a depth z using this equation. If the stopping cross-section $[\varepsilon]$ is used instead of dE/dz, the equation becomes:

$$\Delta E = N.z.[\varepsilon]$$

If the energy difference ΔE is chosen to be the energy width δE of a single channel in the multichannel analyser, then each channel corresponds to a thickness δz:

$$\delta z = \delta E / N.[\varepsilon]$$

The depth resolution in RBS with ^{4}He$^+$ 2.0 MeV ions is better than 10 nm for near-surface analysis and with a system resolution of 15 keV. Depth profiles up to microns can be obtained.

4.3.2.7. Determination of Elemental Concentrations.

Experimental RBS spectra are usually interpreted by means of computer simulation techniques. The best known of these is the RUMP code (Doolittle 1985) which generates theoretical spectra for thick or thin targets bombarded by light projectiles with incident energies up to 4 MeV. Starting from a hypothetical description of the sample (the atomic elemental composition as a function of depth) the interpretation of the data proceeds by iterative simulations of the spectrum and correction of the sample description, then seeking the best simulation among a series of simulations computed with changing parameters.

In the RUMP code, ('Rutherford Universal Manipulation Program') the sample is considered to be a stack of layers, each of uniform composition, and the code calculates an arbitrary density from their elemental composition. All the layers constituting the sample are thus considered to be of homogeneous composition and density, and the total number of incident particles and their solid angle is inserted into the code by the user. The code simultaneously provides an elemental depth profile (which can differ from the initial input description of the sample) as well as theoretical spectra.

An example of this process of data analysis is provided by the work of Yubero *et al.* (2000), who studied the structure of iron oxide thin films prepared at room temperature by ion beam induced chemical vapour deposition. Such films find important applications because of their optical, magnetic, or magneto-optical properties. They were produced by bombardment of a substrate with O_2^+ or $O_2^+ + Ar^+$ mixtures, and Figure 4.15 shows RBS spectra of two iron oxide thin films prepared on a Si substrate by each of these bombardment methods.

Simulation of the RBS experimental spectra by means of the RUMP code has shown that, in both cases, the film composition was Fe_2O_3 (within 5%). The spectra also confirm the sharp character of the substrate/film interface, within the resolution

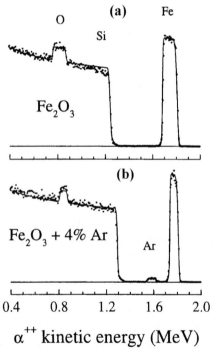

Figure 4.15. RBS characterisation of iron oxide thin films prepared with ion bombardment with (a) O_2^+ or (b) $O_2^+ + Ar^+$ mixture. Dots: raw data. Full lines: RUMP simulations. (Reproduced by permission of Yubero *et al.* 2000.)

limits of the technique, i.e., ≤ 10 nm. The thickness of these two particular films was shown to be 140 nm and 90 nm, which corresponds to 1300 at/cm^2 and 770 at/cm^2 respectively, assuming a density of 5.0 g/cm^3 typical of Fe_2O_3. The RBS spectra also indicate that in the film prepared with the $O_2^+ + Ar^+$ mixture, 4% atomic Ar is homogeneously incorporated within the film.

4.3.2.8. Summary of the Characteristics of RBS

Probe energy	1–3 MeV
Beam diameter	~0.5–1.0 mm (~1 μm with microbeam)
Beam current	~2–20 nA
Analysis time	~5–30 min
Scattering angle	170°
Energy analyser	surface barrier detector 12–15 keV energy resolution (ΔE_{det})
Probing depth	~1–2 μm
Depth resolution	20–30 nm (3–4 nm with tilted targets)
Mass resolution	isotope resolution up to ~40 amu
Sensitivity	10^{-2}–10^{-4} monolayers for heavy surface impurities
	10^{-1}–10^{-2} monolayers for light surface impurities
Accuracy	3–5% (typical)

4.4. HEAVY ION BACKSCATTERING SPECTROSCOPY (HIBS)

HIBS is the same as RBS, except that heavy ions are used instead of He^{++}. It is an ion beam analysis tool patented by the Sandia Corporation of the USA, and was developed to enable the measurement of trace levels of surface contamination on silicon wafers. Metal contamination present in starting material is detrimental to devices, since it results in defects which limit wafer yields and impair circuit operation.

HIBS is capable of measuring an ultra-trace level of surface contamination of $< 1 \times 10^{10}$ atoms/cm^2, by taking advantage of the increased backscattering yield from a low energy C$^+$ beam focused on to a silicon wafer. The near-surface impurities are identified and quantified by measuring the time-of-flight (TOF) of the backscattered C with an appropriate detector array. HIBS is especially useful when examining layers of III–IV and IV–VI semiconductor compounds such as GaAs, InSb and SnTe whose constituents lie in the same row of the periodic table. The similar atomic masses of these constituents are not readily resolved by conventional RBS, and furthermore many of these atomic species possess multiple isotopes which may be introduced features in HIBS spectra which cannot be resolved in RBS spectra of the same sample.

4.4.1 Resolution of Ga and As Peaks in GaAs Layers

The work of Stumborg *et al.* (1998), illustrates the advantages of HIBS over RBS in their study of GaAs layers grown by Molecular Beam Epitaxy (MBE) upon BaF_2 layers. Small cross-sections make RBS inappropriate for the detection of the monolayer scale coverages of materials that occur during the initial stages of MBE growth, and these authors had previously (Stumborg *et al.* 1996) used HIBS to detect successfully sub-monolayer coverages of Ba during the growth of BaF_2 layers on GaAs substrates.

The 1998 paper describes the use of HIBS to separate backscattered signals of Ga and As from thin films of GaAs grown on BaF_2. There is no widely accepted standard reference for values of electronic stopping power (S_e) of heavy ions, and limits the use of HIBS for film thickness measurements. The backscattering cross-section (σ) and kinematic factor (K_m) factor calculations for HIBS are of uncertain validity, so that HIBS data are generally limited to qualitative analysis, except in situations where non-Rutherford correction factors can be neglected.

RBS and HIBS analyses were performed using 2 MeV $^4He^{++}$ ions beams for the former analyses and 12 MeV $^{12}C^{4+}$ ion beams for the latter. Beam currents on target were of the order to 10 and 60 nA for the He and C beams, repectively. Samples were placed in the scattering chamber with target angles ranging from $0°$ to $80°$. Backscattered ions were detected by a solid state surface barrier detector at an angle of $170°$ with respect to the forward direction.

Simulation spectra were generated using parameters that describe the ion beam, target and detector geometry, beam and detector resolution, and sample characteristics. The sample parameters, which include the number of layers and the areal density and atomic composition of each layer were then varied until the simulation conformed to the experimental data. The HIBS spectra were analysed using a modified version of the RBS analysis program.

Figure 4.16 shows the RBS and HIBS spectra of a specimen formed by depositing BaF_2 for 10 min at 650°C; the temperature was then reduced to 575°C and GaAs was deposited for 10 min. The number of counts in the Ga and As signals have been magnified by a factor of 10 for clarity. The HIBS data were analysed first using a 12 MeV ^{12}C beam simulation. This analysis produced a set of beam and sample parameters that gave a best fit to the HIBS data. The beam parameters were then changed to the parameters of a 2 MeV ^4He beam, and the sample parameters were left unchanged. This new set of beam/sample parameters was then used to

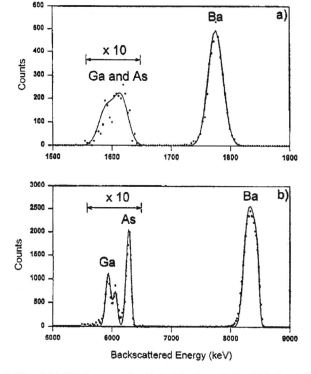

Figure 4.16. (a) RBS and (b) HIBS spectra for GaAs islands on BaF_2 (111). Symbols: ● data points, – simulation. (Reproduced by permission of Stumborg *et al.* 1998.)

generate an RBS simulation. The simulations are for small islands of stoichiometric GaAs on BaF_2. The 10 min deposition time does not provide full coverage of the BaF_2 surface.

The enhanced mass resolution of HIBS is clearly evident in Figure 4.16. The Ga and As signals are clearly resolved, and the two isotopes of Ga are nearly resolved (As is mono-isotopic). This mass separation provides information essential to the determination of the GaAs epilayer composition (which is stoichiometric in Figure 4.16(b)). This is not possible with the overlapping Ga and As signals of Figure 4.16(a), which can be fitted to a range of Ga-to-As ratios.

The mass resolution of HIBS ranges from ~ 2 amu for Fe up to ~ 20 amu for Pb.

4.5. PROTON-INDUCED X-RAY EMISSION (PIXE)

A publication by Johansson *et al.* (1970) over thirty years ago marks the introduction of this technique of particle-induced X-ray emission analysis. They used protons and

Si(Li) spectroscopy, with the capability of simultaneous quantitative analysis of 72 elements ranging from sodium through to uranium in solid, liquid, thin film and aerosol filter samples. The penetrating power of protons allows sampling of depths of several tens of microns, and the beam itself may be focussed, rastered or varied in energy. The use of a proton beam as an excitation source offers several advantages over other X-ray techniques, for example there is a higher rate of data accumulation across the entire spectrum which allows for faster analysis.

Many applications of PIXE take advantage of its ability to cope with very small specimens (0.1–1 mg) – achieving detection limits below 1 ppm in favourable situations. The good overall sensitivities, especially for the lower atomic number elements, is due to a lower Bremsstrahlung background resulting from the deceleration of ejected electrons, in comparison with electron excitation, and the lack of a background continuum compared with X-ray fluorescence analysis (XRF). It may be regarded as a non-destructive technique, and is frequently applied to valuable specimen materials such as works of art or archaeological artefacts.

The arrangement is shown schematically in Figure 4.17.

PIXE is based upon a small particle accelerator providing a beam of protons (or in a few cases helium or even heavier ions) which passes first through a bending magnet having a very stable field and is then stabilized by passing it through a slit. The beam is directed axially down the beam-line using electrostatic and magnetic steering elements, under a typical vacuum of 10^{-6} torr.

The specimen chamber (or target chamber) may contain a number of samples and standard specimens mounted on the same sample holder. Also here are the X-ray detection system and a Faraday cup which monitors the proton current incident

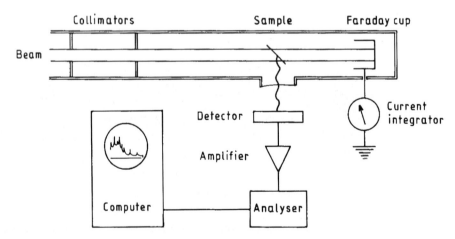

Figure 4.17.   Schematic diagram of the arrangement for PIXE analysis.

upon the specimen. The particle beam passes through the sample, causing the emission of X-rays which are detected using a lithium-drifted silicon detector, which has a good energy resolution and high efficiency in the X-ray energy region 2–20 keV. The resolution is such that one can fully resolve the K_α X-rays of neighbouring elements in the transition element region of the periodic table. The heavy elements are determined by means of the L_α X-rays, whose energy is about the same as the K_α X-rays of the elements mentioned.

The height of a given X-ray peak is a measure of the amount of the corresponding element in the sample. The X-ray production cross-sections are known with good accuracy, the beam current can be measured by, for example, a Faraday cup (Figure 4.1) and the parameters of the experimental set-up are easily determined so that the sample composition can be determined in absolute terms.

Johansson *et al.* (1995) illustrate the detection limits for the above type of specimens in terms of concentrations (ppm). Contours are shown in Figure 4.18 showing the dependence of the detection limit upon the trace element atomic number (Z) and the proton beam energy E_p.

It is apparent that PIXE exhibits its maximum sensitivity or minimum detection limit (MDL) in the two atomic number regions $20 < Z < 35$ and $75 < Z < 85$. These are attained at relatively low proton energies, which implies that small accelerators are most suitable for PIXE with the corresponding benefits in reliability and economics. Analysis times are typically a few minutes in duration. The MDL is very strongly influenced by the nature of the sample, especially if there are strong X-rays from the matrix visible in the spectrum or if the sample is strongly insulating.

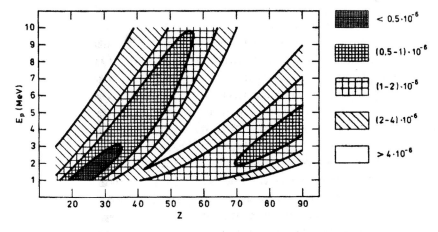

Figure 4.18. Minimum detectable concentrations as a function of atomic number and proton energy for thin organic specimens in a typical PIXE arrangement. (Reproduced by permission of Johansson *et al.* 1995.)

The characteristic X-rays of the lightest elements have such low energy that they are easily absorbed in the sample and in the windows of the chamber and the detector. The practical limit for PIXE analysis is around $Z = 13$ (aluminium), and lighter elements are determined by exploiting various nuclear reactions within the same experimental set-up. One suitable reaction is proton capture leading to gamma-ray emission, PIGE (q.v.), and PIXE and PIGE analyses can be performed simultaneously.

4.5.1   The Ion Beam

UHV is not mandatory in PIXE, and the vacuum shared by the beam-line and specimen chamber is typically $\sim 10^{-6}$ torr. The beam emerging from the accelerator has to be made uniform, while generating the minimum possible X-ray and γ-ray background near the Si(Li) detector, and to this end graphite or tantalum collimators are to be preferred. PIXE chambers are often lined with graphite foil.

The radial intensity of the beam supplied by the accelerator is roughly Gaussian in cross-section, but in PIXE experiments a rectangular profile is required. This can be effected in several possible ways: *quadrupole focussing magnets* with electrostatic deflector plates can produce the requisite spot size without the use of apertures (which are a source of undesired background). Alternatively, a *thin foil* placed 0.5–1 m upstream of the target will cause the protons to undergo small-angle scattering leading to 1–10% of the intensity to be retained at the target – the rest being removed by collimators. A third method is to *defocus the beam* with magnetic quadrupole magnets, which broadens the Gaussian beam profile, and an aperture can select the central, flat-topped portion of the beam intensity.

A small fraction of the available beam current may be selected by means of a diaphragm containing a hole of adjustable diameter in the range 10–100 μm. dimensions a few tens of microns, a technique known as μ-PIXE. By means of magnetic lenses, the beam is focussed so that it gives a demagnified image of the diaphragm on the sample to be analysed. With a demagnification factor of ten, the beam spot on the surface of the sample has a diameter in the range 1–10 μm. With some sacrifice in beam current, even submicrometre spatial resolutions have been achieved.

Under conditions of μ-PIXE, the beam can be scanned across the surface of the specimen and thus provide concentration data as a function of position. Since spectra have to be accumulated at many points on the specimen with this technique, the analysis duration may be hours rather than minutes. Concentration values at different points or concentration profiles for selected elements can be determined by this approach. A detailed account of the technique and its applications is given by F. Watt and G.W. Grime (in Johansson *et al.* 1995, p. 101).

4.5.2 The Specimen Chamber and Detectors

The Si(Li) detector usually views the specimen at either 90° or 135° to the beam direction, either through a thin window in the chamber wall, or for detecting the K X-rays of the light elements below atomic number 20 the detector may be placed within the chamber itself, thus dispensing with the need for a window.

The detector itself may be shielded from background γ-rays by means of an annular shield of W or Pb, and absorbers in the form of appropriate metal foils are placed between the detector and the specimen. These reduce the intensity of the continuum of Bremsstrahlung radiation and also prevent back-scattered protons from entering the detector which would degrade the observed spectrum.

The chamber may also be equipped at 180° to the beam with a (silicon surface barrier) detector for analysis of scattered protons, which provides the option of performing quantitative light element analysis by RBS (q.v.). In certain applications RBS can determine most of the matrix composition and PIXE the trace element contribution.

4.5.3 The Specimen

In the case of thin samples described so far, the absorption of x-rays in the sample is negligible, and it is not necessary to apply any corrections for the slowing down of the particles or the absorption of X-rays.

If 'thick samples' are placed in the specimen chamber for analysis, the particles are slowed down and eventually stopped in the sample, so the calculation of the X-ray yield and their absorption is more complicated. Some objects may be too large to be placed in the specimen chamber, in which case the external beam technique is employed. The particle beam passes through a window at the end of the beam-line into the air where an object of any size (e.g. an archaeological artefact) may be analyzed.

4.5.4 The Spectra

Computer software codes are available to deconvolute PIXE spectra and to calculate peak areas with accuracy, so that absolute amounts of elements present in the specimen may be derived. With a beam of 5 mm diameter incident on a thin organic specimen on a thin backing foil, trace elements can be detected at picogram levels. The x-ray production cross-sections, absorption coefficients and the various experimental parameters are introduced into the computer code employed, so that a print-out of mass values of all the determined elements is obtained.

The advantage of μ-PIXE analysis over the scanning electron microprobe arises from the presence of a strong Bremsstrahlung background in the latter, which tends to mask the characteristic X-ray peaks. There is thus a striking difference in *sensitivity* between the two techniques: the detection limits are of the order of 0.1% for the electron microprobe and 0.001% for μ-PIXE.

In conventional PIXE the beam diameter is a few millimetres, which gives detection limits of the order of 10^{-11} g. With μ-PIXE and a spatial resolution of about 1 μm, detection limits as low as 10^{-16} g can be achieved.

Figure 4.19 shows the μ-PIXE X-ray spectra collected simultaneously from a fragment ($\sim$1 mm in diameter) of an archaeological cobalt-blue glass (Uzonyi *et al.* 2001). This sample contains numerous minor and trace elements from carbon to lead, and the instrument employed both an ultra-thin window (UTW) detector as well as a Be-windowed detector.

These workers employed a new in-vacuum experimental set-up, and it can be seen that the analytical range of the UTW detector is between $\sim$0.2 and 6 keV (which corresponds to elements from C to Mn), whereas that of the Be-windowed detector is above 4 keV, corresponding to elements from Ti to U. A PIXE spectrum is usually quite complicated, and due to interferences between different elements and due to small peaks being hidden by larger ones. Deconvolution by computer using special codes is required in order to carry out the complete analysis of the spectrum.

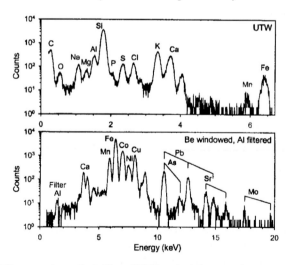

Figure 4.19. PIXE X-ray spectra collected by μ-PIXE analysis in a new in-vacuum experimental set-up from a fragment of cobalt-blue glass. Upper spectra from a UTW detector and lower spectra from a Be-window detector. (Reproduced by permission of Uzonyi *et al.* 2001.)

4.5.5 Applications: Analysis of Air Pollution Samples

A matrix of light elements (e.g. plastic, or organic tissue) containing a large number of trace elements constitutes an ideal sample for PIXE analysis, and air pollution samples provide a good example of the technique.

μ-PIXE can be used to identify and analyse individual particles in aerosol samples down to submicrometre dimensions. Kasahara *et al.* (2001) have used this approach to characterized individual aerosol particles collected during an Asian dust storm in Japan during 1999. Figure 4.20 shows a schematic diagram of the beam scanning and data acquisition system employed.

The sample environment was filled with He gas to prevent the argon X-ray emission from air. Beam scanning, data acquisition, evaluation and the generation of elemental maps were controlled by a computer. Micro-PIXE measurements were performed with a scanning 2.5 MeVH$^+$ microbeam accelerated by the 3 MV single-end accelerator. The beam diameter was 1–2 μm, so that individual particles could be analysed. The beam current was < 100 pA and the irradiation time was about 30–40 min.

The atmospheric aerosols were filtered by a two-stage sampler that classified them into fine (< 1.2 μm) and coarse (> 1.2 μm) fractions, which were then further classified into 13 size ranges between 0.01 to 30 μm. An example of a PIXE spectrum of a coarse particle is illustrated in Figure 4.21.

The large amount of S in the particles suggested that SO$_2$ gas molecules or small sulfur-containing particles condense on to the surface of soil dusts during their transportation from China. Figure 4.22 illustrates an elemental map for Si distribution in coarse particles within a total scanning area of 25 μm × 25 μm. The scale bar shows the peak count of characteristic X-rays by pixel of the scan area.

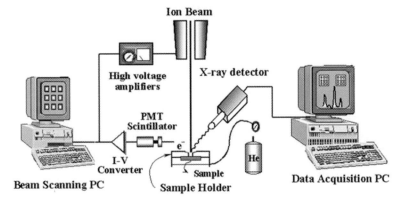

Figure 4.20. Schematic diagram of the micro-PIXE beam scanning and data acquisition (Kasahara *et al.* 2001.)

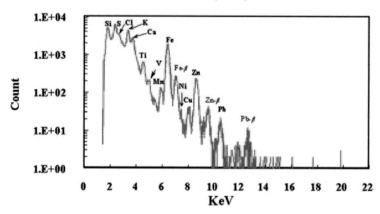

Figure 4.21. Micro-PIXE spectrum of a coarse particle ($> 1.17\,\mu m$) collected during an Asian dust storm. Beam: $2.5\,MeVH^{+}$, current $70\,pA$, irradiation time: $30\,min$. (Kasahara *et al.* 2001.)

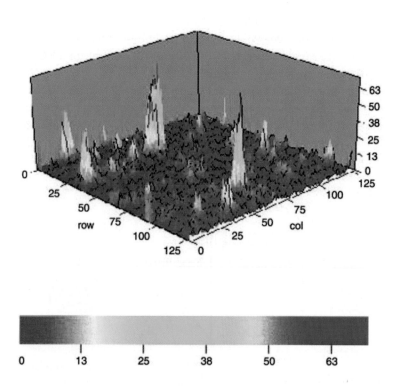

Figure 4.22. Micro-PIXE elemental map for Si taken on coarse particles ($\leq 1.2\,\mu m$) within a total scanning area of $25\,\mu m \times 25\,\mu m$. (Kasahara *et al.* 2001.)

More than 20 Si-containing particles are seen to exist within the $25\,\mu m^2$ area shown. Ca, Fe and Si were found to be the most abundant components in single particles.

4.5.6 The Determination of Elemental Composition and Distribution Profiles

Faiz *et al.* (1996) have applied micro-PIXE analysis to study solute distributions in a single crystal sample of $Y_1Ba_2Cu_3O_{7-\delta}$ high temperature superconductor (YBCO) of dimensions $1.3\,mm \times 1.5\,mm \times 75\,\mu m$. It contained a small secondary crystal overgrowth of dimensions $340 \times 340 \times 100\,\mu m^3$. The interface region between the smaller crystal and the base crystal was covered with a material which appeared to be residual flux. The instrument employed a 2.5 MeV focused proton beam of about $4\,\mu m$ resolution, which could scan an area of $500 \times 500\,\mu m^2$ on the sample surface. The microbeam current was kept low (typically about 30 pA) to avoid any damage to the sample.

An average compositional spectrum as well as two-dimensional element distribution maps were acquired simultaneously from each scanned area. Line spectra could also be obtained by allowing the microbeam to scan along a line across the crystal face. The authors examined the base crystal, the smaller crystal as well as the interface region.

Figure 4.23 is a typical PIXE spectrum from the base crystal, showing the presence of Au and traces of Cr and Fe as impurities, in addition to Y, Ba and Cu.

The source of Au was most likely the gold crucible generally used for crystal growing. A comparative analysis of the crystal energy dispersive spectrometry using a SEM (q.v.) identified the primary elements, plus a very weak peak for Au and no indication of Cr and Fe traces. This demonstrates the advantage of PIXE over SEM in terms of sensitivity for trace element detection.

Figure 4.24 shows the spatial distribution maps obtained from Y, Ba, Cu and Au by micro-PIXE.

The elements are seen to be homogeneously distributed over the base crystal as well as on the smaller crystal. There are strong differences apparent in the elemental distribution patterns in the interface region between the two crystals, although this region contains the same YBCO compounds as uncrystallized flux. Ba and Cu have higher concentrations here than in either of the two crystals, because of the use of an excess of BaO and CuO in the starting compounds when the crystals were grown from the melt. These differences in the elemental distributions can be seen more clearly in Figure 4.25, which is a line scan generated by scanning the sample along a line running across parts of the two crystals and the interface region.

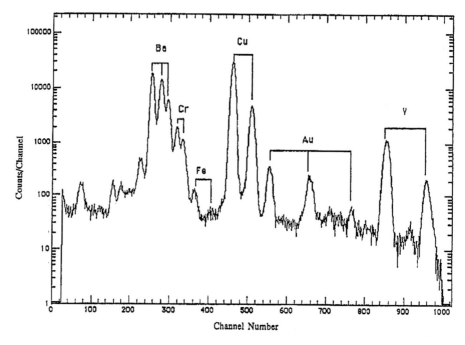

Figure 4.23. A typical PIXE spectrum from the scanned area on the base YBCO crystal.
(Reproduced by permission of Faiz *et al.* 1996.)

The scanning line was chosen to cross the interface region twice, and the concentration differences of the elements in the three regions were confirmed by Faiz *et al.* by PIXE measurements on fixed spots in these areas.

4.6. PROTON INDUCED GAMMA-RAY EMISSION (PIGE)

In contrast to PIXE and RBS, where forces are respectively electromagnetic and electrostatic, this kind of microanalysis uses low range nuclear forces. The analysis is based on the detection of the γ-rays emitted from nuclei that are in an excited state following a charged particle induced nuclear reaction.

The conservation of energy during the nuclear reaction analysis experiment may be expressed:

$$E_a + Q_i = E_b + E_z + E_\gamma$$

where E_a, E_b and E_z are the kinetic energies of the incident particle nucleus, the emitted particle and the residual particle nucleus respectively, Q_i is the energy

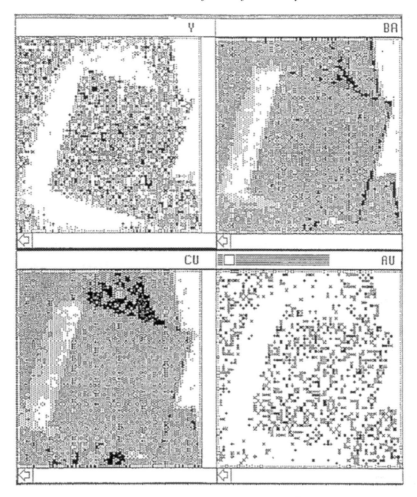

Figure 4.24. Elemental distribution maps from an area of $500 \times 500 \, \mu m^2$ covering the smaller crystal, the interface region and part of the base crystal of YBCO. (Reproduced by permission of Faiz *et al.* 1966.)

corresponding to the i excited state of the residual nucelus and E_γ is the energy of the emitted gamma ray.

The energy of the γ-rays is indicative of the isotope present, and the intensity of the γ-rays is a measure of the concentration of the isotope in the sample. The limitation of this method is that, in order to have a nuclear reaction, the repulsive Coulomb barrier has to be overcome. For incident particles of energy up to 3 MeV, the only accessible elements are the light elements with $Z < 15$: the cross-sections of the remaining elements become rapidly negligible.

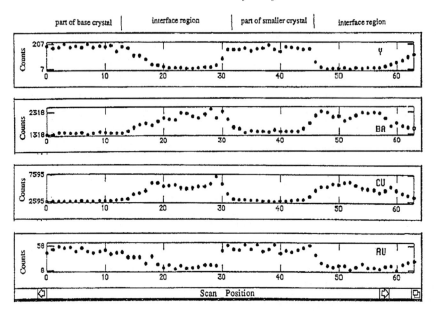

Figure 4.25. Elemental distribution spectra along a line running across parts of the two crystals of YBCO and the interface region. (Reproduced by permission of Faiz *et al.* 1996.)

PIGE is a rapid, non-destructive technique that is employed in the analysis of light elements such as *lithium* (10–100 ppm limit of detection), *boron* (500–1000 ppm limit of detection), and *fluorine* (1–10 ppm limit of detection), which are often difficult to determine by other analytical means. Because the technique is based upon specific nuclear reactions, the sensitivity of PIGE varies greatly from isotope to isotope, and this non-uniformity of sensitivity has limited its widespread use as a complementary technique to micro-PIXE.

4.6.1 *Instrumentation*

Various additional detectors are often incorporated in a PIXE system (q.v.) to take advantage of other particle-induced processes. An intrinsic Ge(Li) or NaI(Ti) detector is often used to conduct simultaneous PIGE analysis. The latter detector has a good efficiency but a bad resolution, while Ge(Li) detectors has a poor efficiency but an excellent resolution.

4.6.2 *Applications*

One typical example of microbeam PIGE analysis is that due to Tadić *et al.* (2000) who studied the distribution of Li and F in a series of gel polymer electrolyte samples containing lithium salt solution, which are used in lithium batteries. The physical chemistry of these gel electrolytes is still unclear, and it is necessary to determine the Li and F distribution in order to improve the polymer preparation procedure, as well as to improve its characteristics as separator in polymer batteries.

Tadić *et al.* studied the polymer poly-vynilidene fluoride/hexa-fluoropropylene ('PVdF/HFP') containing lithium salt solution in Ethylene carbonate/diethylene carbonate ('EC/DEC'). In order to understand better the effect of anion size in the electrolyte, two Li salts were compared, namely $LiN(CF_3SO_2)_2$ (termed 'Liimide' by the authors) and $LiN(C_2F_5SO_2)_2$ (termed 'Libeti').

Microbeam scanning of the sample cross-section was performed with an external microbeam (in air), using a focused 4 MeV proton beam and a 50 μm thick Kapton foil at the vacuum–air interface, with a 5 mm diameter beam exit hole. The 2 mm thick slice of gel polymer sample was placed less than 100 μm from the exit foil, with the cross-section facing the Kapton foil. A HPGe γ-ray detector was placed just behind the sample in order to achieve as large as possible detector solid angle. The ion current was kept below 100 pA in order to minimize damage to the sample.

In the present experiment, the PIGE analysis used the Li γ-ray of energy $E_\gamma = 478$ keV from $^7Li(p, p_1\gamma)$ ^7Li reaction and the F γ-ray of energy $E_\gamma = 197$ keV from ^{19}F (p, $p_1\gamma$) ^{19}F reaction.

Some experimental results for each of the two Li salts are shown in Figure 4.26 as plots of γ-ray yield per channel. The yield of the F γ-ray is divided by 40, and a 3-point smoothing procedure was applied to the yield of both γ-rays. These γ-ray yields are the sums of yields from the sample cross-section surface to a depth of 123 μm, where about 50% of yield of both γ-rays is created within 53 μm from the surface. The data of Figure 4.26 therefore represent the depth average of Li and F distribution within the same depth. The authors conclude that the Li distributions in the studied samples seem to be homogeneous down to 100 μm. Local decreases in the F γ-ray yield are accompanied by increases in the Li γ-ray yield, and this is due to the presence of gaps in the PVdF/HEP polymer leading to local acculumations of EC/DEC solution.

In a later paper, Tadić *et al.* (2001) employed proton microbeam scans over the sample cross-section in order to establish simultaneously the spatial distribution of Li and other chemical species in the vicinity of the interface between the gel-polymer and the anode or cathode. Figure 4.27 show the result of a scan of $600 \times 800\ \mu m^2$ across the interface between the cathode (upper part of the map) and the gel-polymer (the lower part).

As in the data of Figure 4.26, it is seen that the local decrease in F γ-ray yield relate to an increase in Li γ-ray yield, due to voids in the spinel cathode that relate to local accumulations of gel-polymer. The authors conclude that PIGE and PIXE

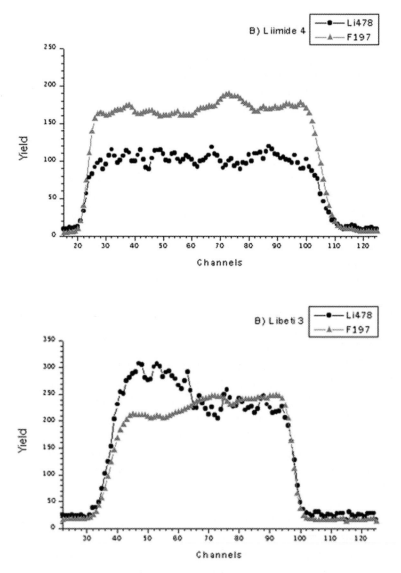

Figure 4.26. Yield distributions of 478 keV Li γ-ray and 197 keV F γ-ray for 4 MeV protons in gel polymer samples containing different Li salts. The F γ-ray yield is divided by 40. (After Tadić *et al.* 2000.)

techniques provide simple and non-destructive methods for the estimation of all elements of interest for Li-ion battery application. They suggest that similar measurements could be applied in future to monitor the elemental distribution changes at cathode/gel-polymer interfaces during the charge/discharge cycles, which may improve present numerical models of the Li-ion battery operation.

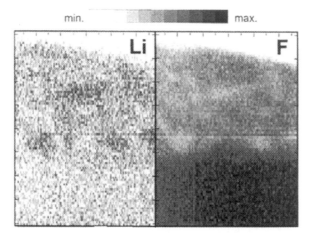

Figure 4.27. Li and F distribution near spinel $LiMn_2O_2$ cathode/gel-polymer interface: PIGE spectroscopy of proton microbeam scan of $600 \times 800 \, \mu m^2$. (After Tadić *et al.* 2001.)

4.7. ELASTIC RECOIL DETECTION ANALYSIS (ERDA)

This technique of *surface analysis* employs a beam of positive ions to irradiate the near-surface region of a target. Although most of the incident particles are elastically scattered by the target atoms, a small number of incident ions are able to induce nuclear reactions on isotopes of light elements situated within the target. The energy transferred during such collisions can be large enough to cause the target nucleus to recoil from the target surface, and ERDA consists of detecting recoiling nuclei in order to acquire information about the composition of the target.

In *transmission* ERDA the beam of monoenergetic ions is directed perpendicularly to the surface of the target. If the target specimen is a thin foil, most of the incident particles and the recoils emerge at the back surface of the target. If it is a thick sample, only those particles with sufficient energy can reach the back surface, and in both situations the recoils emerging at a given angle can be detected.

In *reflection* ERDA the ion beam impinges on to the specimen at grazing incidence, and the recoils emerge from the front surface, generally also at a grazing angle.

The ions originating from the incident beam are often absorbed by means of a foil placed in front of the detector. There are a number of detection systems which may be employed to identify the various recoil species, including electric or magnetic mass spectrometers, and time-of-flight detectors. Qualitative analysis is achieved by characterising the atomic masses of the recoiling nuclei through the energy transfer associated with the collision. Quantitative analysis requires a determination of the *recoil yield*, from a knowledge of the scattering cross-section that describes the probability that scattering events occur.

Tirira *et al.* (1996) give a detailed account of both the theory and application of the technique to the determination of hydrogen in solids, to which reference may be made by the interested reader.

As with RBS, one may determine the composition of the target as a function of depth. In the case of ERDA, the energy of detected recoil nuclei depends upon the energy loss experienced by the impinging ions before collision, as well as that by the recoiling nuclei afterwards. This energy loss depends on the *stopping power* of the target for both projectile and recoils, and the energy spread due to energy loss fluctuations (*energy straggling*) has also to be taken into account.

From a specimen surface of typical area $5 \times 15\,mm^2$, ERDA may thus provide a non-destructive determination of hydrogen isotopes and their depth profiles in polymers and other solids with a sensitivity of 0.01 atomic %, a depth resolution of about 10 nm close to the surface, but decreasing with increasing depth. The maximum depth of analysis is about 1 μm (depending on the material). Measurements of other light elements ($Z < 9$) are also possible if heavy ions such as Cl or Au are used.

At present, more than 50 laboratories in the world currently use ERDA. Tirira *et al.* (1996) give an overview of the activities of the various groups involved in this work, quoting one representative paper from each group they list in order to illustrate its contribution to ERDA investigations. As one example of an ERDA setup, we will quote from Bohne *et al.* (1998), who describe the time-of-flight EDRA (TOF-ERDA) spectrometer at the Hahn-Meitner-Institut Berlin. Many different projectile ions from hydrogen to xenon with variable energies up to several MeV/u are available at this laboratory. High resolving power with respect to mass and energy is possible when TOF is used, because accurate time calibration is easier to perform than energy calibration, and this can be done using the very precisely known velocity of α-particles from radioactive sources.

A schematic diagram of the Berlin TOF-ERDA setup is shown in Figure 4.28.

An outline of the operation of the systems is as follows: The high-energy heavy ion beam is focussed on to a 6-position sample holder in the centre of a 40 cm diameter target chamber. The angle of the sample surface to the beam axis is chosen normally to be 70°, because the standard detection angle of the recoils is 40° relative to the beam direction. The detector system may also be mounted in the smaller triangular shaped part of the chamber, which can be connected to the low energy beam line. Experiments may be performed with low and high energetic beams simultaneously.

The TOF energy telescope consists of two channel-plate timing detectors followed by a silicon energy detector. A simple 'short' TOF spectrometer is also

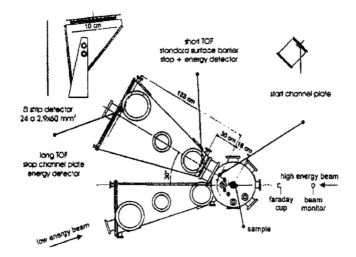

Figure 4.28. Schematic diagram of the Berlin TOF-ERDA setup. (Reproduced by permission of Bohne *et al.* 1998.)

installed, and a comparison of the corresponding spectra recorded with both telescopes, the detection efficiency of the 'long' spectrometer is determined. Calibration measurements for energy and time, measurements of energy loss and time resolution have to be performed using well-characterised α-particle sources and recoil material.

4.7.1 Hydrogen Determination in Solids by ERDA

There is a wide range of application fields employing hydrogen isotope determination by ERDA, including thin films, elemental transport phenomena near interfaces, materials for microelectronics, and polymers. Polymers are amongst the most hydrogen-rich media that can be found, and numerous applications of the technique have been applied in polymer science. As an example, we will describe the measurement of hydrogen and deuterium profiles occurring in polymer blends.

Bruder and Brenn (1992) studied the spinodal decomposition in thin films of a blend of deuterated polystyrene (dPS) and poly(styrene-co-4-bromostyrene) (PBr_xS) by TOF-ERDA. They examined the effect of different substrates on the decomposition process. In one series of experiments, a solution of the polymers in toluene was spread on a silicon wafer to form a film of thickness 550 nm which was then heated in vacuum at 180°C for various times.

After quenching to ambient temperature the laterally averaged volume fraction versus depth profile was measured by TOF-ERDA, and Figure 4.29 shows a series of dPS profiles for increasing annealing times. After 10 min annealing at 180°C,

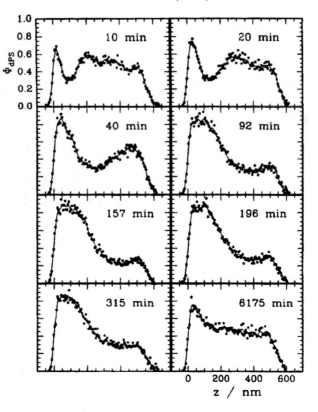

Figure 4.29. Volume fraction versus depth profiles of dPS normal to the film surface after various times of annealing at 180°C. (Bruder and Brenn 1992.)

a dPS-rich surface layer has been established on the vacuum side, followed by oscillations as expected for spinodal decomposition. With increasing annealing time this structure coarsens, and after 157 min a macroscopic surface layer can be identified with a maximum extension of ~ 220 nm, and the layer near the silicon surface has a mean composition of 0.23 for longer times it is seen that this structure is not stable, but decays continuously.

The lateral structure was studies by optical microscopy, and contrast arises from the fact that after irradiation of the samples with the ion beam in the TOF-ERDA apparatus, the PBr_xS-rich structures appear darker than those rich in dPS, giving an excellent contrast (Figure 4.30).

Lateral structures were observed after 40 min of annealing, when typical spinodally decomposed structures were apparent. After 315 min the mean diameter of the dPS particles has reached the film thickness, and for larger times, two-dimensional domain growth takes place.

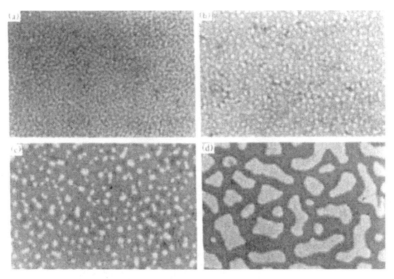

Figure 4.30. Optical micrographs showing concentration fluctuations in the film plane of the specimens of Figure 4.29. The bright regions are those rich in dPS and the dark regions are rich in PBr_xS. The base side of the photographs $= 35\,\mu m$. (a) 40 min (b) 157 min (c) 315 min and (d) 6175 min. (Bruder and Brenn 1992.)

4.7.2 Heavy-Ion ERDA for Measurement of Cr Dopant in Thin Films of β-FeSi₂

Bohne *et al.* (2000) have employed heavy-ion TOF-ERDA to analyse Cr when used as a p-type dopant in $FeSi_2$ films. Cr and Fe have similar masses, and these authors show that heavy-ion ERDA is a powerful tool in this context, being superior to RBS.

These workers grew Cr-doped $FeSi_2$ films on thickness 200–350 nm on Si substrates, and ERDA measurements were performed with a mass and energy dispersive TOF setup as already described. The projectiles were 140–250 MeV ^{129}Xe ions, and the flight path of 123 cm and a time resolution of 135 ps provided a sufficiently high mass resolution to discriminate the Cr recoils from those of Fe. To obtain the sample composition, the energy spectra were simulated by an appropriate computer program.

A scatterplot of energy for an $FeSi_2$ film with only 0.2 at% Cr as dopant is shown in Figure 4.31. In the TOF signal, t_0 includes all the signal delays of the whole set-up. The more abundant isotopes of all contributing elements are seen to be clearly separated.

The integrated mass distribution of the metallic film components is shown in Figure 4.32, and a concentration of 0.20 ± 0.03 at% is extracted. For higher concentrations of dopant the precision is much better. Figure 4.33 depicts the results of the dopant analysis by Bohne *et al.* (2000), compared with the nominal concentration of Cr dopant added during film growth. Standard RBS and heavy-ion

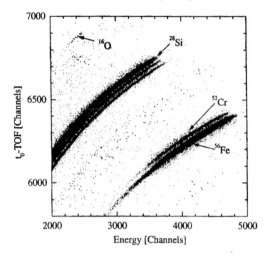

Figure 4.31.   (t_0-TOF) versus energy spectrum of 295 nm Cr doped $FeSi_2$ on Si measured at $40°$ with 230 MeV ^{129}Xe ions (Bohne *et al.* 2000).

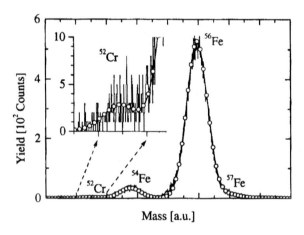

Figure 4.32.   Integrated mass distribution of metallic components for the sample shown in Figure 4.31. (Bohne *et al.* 2000.)

RBS experiments were also carried out by these workers, and it is apparent that the latter data scatter considerably and deviate significantly from those of ERDA.

This arises from inhomogeneous Cr depth profiles, especially at higher Cr content, where ERDA energy spectra clearly displayed a decreasing Cr content towards the film surface. In this concentration range the ERDA results fit a regression line with a slope < 1 crossing the line of slope 1 at an added Cr concentration of about 1.6 at%. The authors suggest this arises because the silicides of Fe and Cr exhibit a low

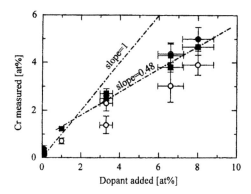

Figure 4.33. Measured concentration of Cr versus concentration of Cr added during film growth for β-FeSi$_2$ films doped with Cr. ■: ERDA; ●: heavy-ion RBS; ○; standard RBS. (Bohne *et al.* 2000.)

miscibility limit (because of incompatible lattice types), so that only a fraction of the offered Cr is incorporated for higher concentrations. The crossing point of the two lines in Figure 4.33 is interpreted as the solubility limit of Cr in β-FeSi$_2$, and the inhomogeneous dopant depth profiles at higher Cr content can be explained by a reduced capability of accommodating Cr at regular lattice sites in the growing FeSi$_2$ film.

4.8. NUCLEAR REACTION ANALYSIS (NRA)

An article by Demortier (2000) gives a more detailed account of the principles of NRA than that which appears below, and to which reference may be made for further information. An incoming ion beam enters the surface of the material and, in the range of energies used, the stopping process is mainly due to interaction of the beam with atomic electron shells (electronic stopping). At a certain depth the incoming particle can undergo a nuclear reaction with atoms of the target, and from a measurement of the energy distribution of the reaction products it is possible to calculate the depth distribution of the atoms that react with the incoming ions.

As nuclear reactions are isotope specific, NRA may be used, for example, to distinguish the distribution of binary blends of polymers in a polymer film, where one of the polymers is labelled with deuterium. The depth distribution of the deuterium atoms can be established and hence that of the labelled polymers.

4.8.1 Symbols Employed in Describing Nuclear Reactions

NRA is concerned with prompt reactions leading to a transmutation of the target atom, but excluding RBS, ERDA and that part of PIGE where the target nucleus does not change during the interaction. The concise form of a nuclear reaction as expressed by nuclear physicists is:

$$A(a, b)B \tag{1}$$

where A is the target nucleus, a the incident projectile, b the emitted particles or photons, and B the residual nucleus which remains in the target and cannot be detected. For example, the capture of a proton in an ^{27}Al nucleus leading to the formation of an excited ^{28}Si which immediately decays to a stable ^{28}Si by the emission of γ-rays would appear as:

$$^{27}\text{Al}(p, \gamma)^{28}\text{Si}$$

The spectroscopy of the γ-rays may be used for analysis.

A list of selected reactions used for the study of light isotopes in a material (atomic number ≤ 20) is given by Demortier (2000). The Q value of a nuclear reaction represents the difference between the total rest mass of the interacting particles and those of emitted ones:

$$Q = (m_A + m_a - m_B - m_b)c^2$$

where c is the velocity of light in vacuum. A positive Q value characterises a nuclear reaction in which some mass is converted into kinetic energy of emitted particles, so that:

$$Q = E_b + E_B - E_a$$

4.8.2 Resonant Energies (E_{ar}) and Depth Profiling

Owing to competing phase shifts in the quantum mechanical behaviour of nuclear reactions, the cross-section may be very low in the vicinity of a resonant energy. These discontinuities in cross-section values induced by a on A give rise to the possibility of *depth profiling* of A nuclei, since signals induced in the detector by b or c particles will be particularly intense when arising from those regions in the target where $E_a = E_{ar}$.

If the target is irradiated with particles of energy greater than E_{ar}, there will be a defined depth below the surface where E_a reaches $E_{ar.}$ The corresponding E_b values may be calculated, and corrected for the energy loss in the outgoing direction on its way to the detector.

4.8.3   Detectors

Solid-state detectors consisting of p-n junctions made with silicon crystal constitute particle detectors for protons, deuterons, ^{3}He and α-particles. The energy of the detected particle is converted into pairs of electrons and holes – the number of pairs being proportional to the deposited energy. For the highest energies, a succession of detectors mounted as a telescope may be necessary, and the associated electronics is designed to collect coincidentally the signals produced in the successive detectors. Measurement of the time-of-flight of the particle may also be used to identify the mass of the particle.

4.8.4   Examples of Application

(1) The data of Zink *et al.* (1998) illustrate the measurement by NRA of near-surface composition profiles in isotopically labelled polymer blends. If a mixture of polymers is adjacent to a phase interface (e.g. a solid or an air surface), often one of the components is preferentially attracted to the surface and will segregate to it, and this phenomenon will influence the tribological behaviour the interface (lubrication, wear and adhesion).

Zink *et al.* used a blend of polystyrene (hPS) and its deuterated counterpart (dPS), both of molecular weight 1.95×10^6 (abbreviated 1.95 M). The average volume fraction (Φ_{dPS}) of deuterated polystyrene was 30%. The polymers were dissolved in toluene and spin cast on thin silicon wafers (about 10×10 mm), the resulting film thickness being about 300 nm. The samples were annealed at 245°C for 8 days, and the measurement of the resulting depth profiles was conducted by NRA using a monoenergetic 700 keV ^{3}He beam. The nuclear reaction employed can be written:

$$D(^3He.p)^4He$$

with a Q value of 18.352 MeV.

The energy of the detected reaction products after leaving the sample can be related to the depth at which the reaction took place through the known energy loss rate of the incoming and outgoing particles as they traverse the sample.

The beam impinged upon the sample at a grazing angle of 8°, and the protons generated by the nuclear reaction were detected in backscattering geometry at an angle of 176° to the incoming beam. The resulting resolution (Gaussian half-width) at the sample surface was 5.5 nm, and Figure 4.34 shows the NRA profiles of the dPS. The experimental points are shown together with the calculated profile.

It is clear that the measured surface enrichment profiles agree well with the theoretical predictions.

(2) Depth profiles of ultrashallow implanted P in silicon have been determined by Kobayashi and Gibson (1999) using NRA. Quantitative analysis of this type is

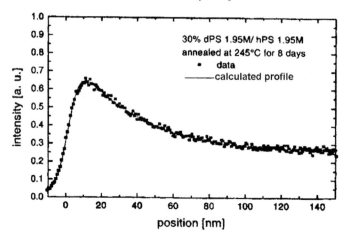

Figure 4.34.   NRA depth profile of the dPS 1.95M/hPS 1.95 M polymer blend (filled squares). The drawn
line is the caculated profile. (Reproduced by permission of Zink *et al.* 1998.)

a critical issue for current Si semiconductor technology, since implanted P can
accumulate near the SiO_2/Si interface due to transient diffusion and anomalous
segregation during thermal annealing. This can cause serious dose loss. Their
specimens contained phosphorus ions implanted into Si (100) substrates at room
temperature at a dose of $3 \times 10^{14} \, cm^{-2}$ in the energy range 5–80 keV.

NRA appears to have better reproducibility than SIMS in the determination of
ultrashallow dopant profiles, and Kobayashi and Gibson employed the $^{31}P(\alpha, p)^{34}S$
reaction and identified the resulting $\sim 2.3 \, MeV$ protons by means of a $200 \, mm^2$
silicon surface barrier detector placed at 171° from the beam direction. A 60 μm thick
Mylar film was placed in front of the detector to eliminate back-scattered alpha
particles, thus providing well-isolated, background-free proton signals. The beam
current was 50 nA and the beam spot size 4 mm × 4 mm; negligible sample damage
by ion beam bombardment was encountered.

Figure 4.35 shows the proton yields of the P implanted Si samples as a function of
incident α energy for four P implant energies.

Gaussian-shaped depth profiles of P with three parameters of maximum
concentration (C_{max}), projected range (R_p) and range straggling (ΔR_p). The energy
loss (dE/dx) and energy straggling (Ω: square root of the variance) of the α beam in
the Si layer were taken into account:

$$dE/dx = 0.177 \, keV/nm$$

$$\Omega = 0.27x\sqrt{t} \, (nm)$$

Optimum values of C_{max}, R_p and ΔR_p were obtained by curve fitting. The accuracy
of R_p is approximately ± 3 nm and is mainly determined by the accuracy of E_{beam}

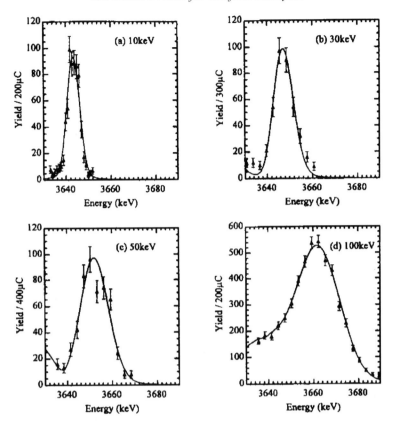

Figure 4.35. Proton yield as a function of incident α energy for P implanted Si samples at implant energies and doses of (a) 10 keV, 3×10^{14} cm^{-2}, (b) 30 keV, 3×10^{14} cm^{-2}, (c) 50 keV, 3×10^{14} cm^{-2}, and (d) 100 keV, 5×10^{15} cm^{-2}. The solid lines are the results of computer simulation. (Reproduced by permission of Kobayashi and Gibson (1999)).

(± 0.5 keV). Figure 4.36 shows a comparison of the experimental results of R_p with computer calculations. The depth resolution of the depth profiles are ~ 7 nm at the surface and ~ 13 nm at a depth of 50 nm; these are determined by ΔE and Ω.

The good agreement indicates that this method is reliable even for ultrashallow depth profiling.

4.9. CHARGED PARTICLE ACTIVATION ANALYSIS (CPAA)

In CPAA, the incident charged particle induces nuclear reactions which produce radionuclides, and the characteristic decay radiation of the latter is measured. Qualitative analysis of the radionuclide is achieved by measuring its energy and/or

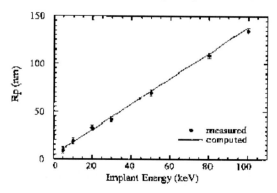

Figure 4.36. Comparison of experimental results with calculated values of projected ranges (R_p) as a function of implant energy for P in Si. (Reproduced by permission of Kobayshi and Gibson 1999.)

half-life, and quantitative analysis is performed by measuring the number of particles or photons emitted, i.e. the radioactivity. The CPs lose their energy when penetrating the sample, and are stopped at a depth known as the *range*. The analysed surface layer is roughly equal to the range, and is typically 0.1 to 1 mm.

CPAA may be employed to determine trace element concentrations in bulk solid material, but its importance in our present context is that it permits the characterization of a thin surface layer, i.e. the mass of the analyte element per surface unit, with a good detection limit and outstanding accuracy. For example the composition of a surface layer (or foil) of known thickness can be determined, or, conversely, the thickness of a surface layer of known concentration. Depth profiling or scanning is not possible, and a disadvantage of the method is that heating occurs during irradiation. It is also not possible to discriminate between different oxidation states of the analyte element or between different compounds.

De Neve *et al.* (2000) have carried out a feasibility study to investigate the possibilities and limitations of CPAA as a thin layer characterization method (rather than for the determination of elemental concentrations in bulk samples). The required experimental conditions are (a) that the surface layer containing the analyzed element is thinner (1 μm or less) than the *range* of the charged particles used and (b) that the substrate (i.e. the layer on which the thin layer is deposited) does not contain the element(s) to be analysed.

4.9.1   *The Partial Mass Thickness*

These authors observe that either 'thick' standards may be used, in which the charged particle beam is completely stopped, i.e. $D \geq R$), or 'thin' standards where $D \ll R$. The partial mass thickness $c_x D_x$ (g cm^{-2}) of a thin layer for a given element is given by Strijckmans (1997) for 'thick' standards as:

$$c_x D_x = c_s \frac{A_x}{A_s} \frac{I_s}{I_x} \frac{(1 - e^{-\lambda t_{i,s}})}{(1 - e^{-\lambda t_{i,x}})} \int_{E_t}^{E_i} \frac{\sigma(E)/\sigma_0}{S_s(E)} dE \qquad (1)$$

$$c_x D_x = c_s \frac{A_x}{A_s} \frac{I_s}{I_x} \frac{(1 - e^{-\lambda t_{i,s}})}{(1 - e^{-\lambda t_{i,x}})} [R_s(E_i) - R_s(E_t)] \qquad (2)$$

For 'thin' standards, the following equation is given:

$$c_x D_x = c_s D_s \frac{A_x}{A_s} \frac{I_s}{I_x} \frac{(1 - e^{-\lambda t_{i,s}})}{(1 - e^{-\lambda t_{i,x}})} \qquad (3)$$

Here, c_x and c_s are the element concentrations (g g^{-1}) in the sample and standard respectively; D_x and D_s are the total mass thickness (g cm^{-2}) of the sample and standard, respectively; A_x and A_s are the activity at the end of the irradiation (counts s^{-1}) for the sample and standard respectively; I_x and I_s are the beam intensity (particles s^{-1}) for sample and standard respectively; t_{ix} and $t_{i,s}$ are the irradiation time (s) for sample and standard respectively; λ is the decay constant of the induced radionuclide (s^{-1}); E_i and E_t are the incident energy of the charged particles and threshold energy of the nuclear reaction, respectively; R_s is the mass range of the charged particles used in the standard (g cm^{-2}); $S_s(E)$ is the mass stopping power of the standard for the charged particles used (MeV g^{-1} cm^2); $\sigma(E)$ is the cross-section of the nuclear reaction (cm^2); σ_0 is the cross-section at the incident energy (cm^2).

4.9.2 Experimental Conditions

De Neve (2000) analysed thin layers of Al, AlO$_x$, TiO$_x$, the high temperature superconductor Yba$_2$Cu$_3$O$_{6+\delta}$ ('YBCO') and Y$_2$O$_3$-stabilized ZrO$_2$ ('YSZ') on different substrates, establishing thereby the accuracy, and detection limits and the precision of the method. The thin layers analysed were made by sputtering deposition upon a polymer substrate in the case of the first three materials listed – the other two were sputtered on Ta substrates. Metal foils were used as standards, and the samples and standards were placed in a water-cooled target holder and irradiated under a He atmosphere with proton (p) and deuteron (d) beams from a cyclotron.

The irradiation and measurement conditions for the analyses are given in the Table 4.2.

This shows the cyclotron energy used in each case, the different foils inserted in front of the samples inorder to monitor the beam intensity and to decrease the

Table 4.2. Irradiation and measurement conditions for the analyses.

Sample	E_{cyclo} MeV	Foils	E_{eff} MeV	I (nA)	t_i	t_c	t_m
Al/AlO$_x$	d 10	Ni (12.5 µm) Ni (50 µm)	5.7	400–500	30 s	4–5 min	5 min
TiO$_x$	p 15	Ni (12.5 µm) Al (3 × 100 µm)	12.0	300	40 min	3 d	24 h
YBCO: Cu and Y	p 15	Ni (12.5 µm) Al (2 × 105µm)	12.7	1000	5 min	Cu: 20 min Y: 1 d	20–30 min 30 min
YBCO: Ba	p 15	Ni (12.5 µm) Al (3 × 105 µm)	12.0	1000	30 min	4 h	2 h
YSZ	p 15	Ni (12.5 µm) Al (100 + 50 + 20 µm)	13.0	1000	10 min	5 h	20 min

Table 4.3. Results of the TiO$_x$ layers obtained via CPAA.

Sample	mean ($\mu g\,cm^{-2}$)	mean (nm)
A	1.0721 (0.0072)	2.517 (0.017)
B	1.1565 (0.0092)	2.715 (0.022)
C	1.1520 (0.0085)	2.704 (0.020)
D	1.125 (0.018)	2.641 (0.043)
E	1.062 (0.011)	2.493 (0.027)
F	7.425 (0.042)	17.430 (0.010)
G	11.835 (0.021)	27.782 (0.050)

incident cyclotron energy to the effective incident energy (E_{eff}), the beam intensity I and the irradiation time, t_i. The cooling time, t_c, and the measuring time, t_m are also given.

4.9.3 The Thickness of Ti$_x$ Layers

To determine the thickness of a thin layer, the composition and the density of the sputtered layer should be exactly known. The partial mass thickness $c_x D_x$ must be divided by c_x (the concentration of the analyte in the thin layer) to determine the total mass thickness D_x. Division of D_x by the density of the thin layer gives the thickness of the thin layer expressed in units of length. In their thickness calculations, De Neve *et al.* (2000) assumed their TiO$_x$ layers to be TiO$_2$ with a density of 4.26 g cm^{-3}.

Each sample was analysed three times, and their results shown Table 4.3 indicate that such layers with a thickness of a few nanometres can be determined with a precision of 2% or better. The standard deviations are given in parentheses.

Results obtained by CPAA for composition and partial mass thickness have been shown to be consistent with the results obtained via other analytical methods. The main advantage of the use of CPAA as a surface characterization method are its purely instrumental character, requiring no sample preparation, its high accuracy, and its low detection limits.

CPAA is too complex to be used as a routine method itself, but it is likely to have particular value in calibrating standard samples for other routine methods.

REFERENCES

Andersen, C.A. & Hinthorne, J.R. (1973) *Anal. Chem.*, **45**, 1421.
Bohne, W., Röhrich, J. & Röschert, G. (1998) *Nuclear Instruments and Methods in Physics Research B*, **136–138**, 633–637.
Bohne, W., Reinsperger, G.-U., Röhrich, J., Röschert, G., Selle, B. & Stauß, P. (2000) *Nuclear Instruments and Methods in Physics Research B*, **161–163**, 467–470.
Bohr, N. (1953) *MAT. Fys. Medd. Dan. Vid. Selsk.*, **27**, 15.
Bruder, F. & Brenn R. (1992) *Physical Review Letters*, **69**, 624.
Chu, W.K., Mayer, J.W. & Nicolet, M-A. (1978) *Backscattering Spectrometry*, Academic Press.
Demortier, G. (2000) Nuclear Reaction Analysis in '*Encyclopedia of Analytical Chemistry*', John Wiley & Sons, Chichester.
De Neve, K., Strijckmans, K., & Dams, R. (2000) *Anal. Chem.*, **72**, 2814–2820.
Doolittle, L.R. (1985) *Nucl. Instrum. Methods Phys. Res. Sect. B*, **9**, 344.
Faiz, M., Ahmed, M. & Al-Ohali, M.A. (1996) *Nucl. Instr. Meth. B*, **114**, 138–142.
Grant, W.A. (1989) in *Methods of Surface Analysis*, J.M. Walls, Ed. CUP, Cambridge p. 299.
Grime, G.W., Watt, F. & Chapman, J.R. (1987) *Nuclear Instruments and Methods in Physics Research B*, **22**, 109–114
Johansson, T.B, Akselsson K.R. & Johansson, S.A.E. (1970) *Nucl. Instr. Meth.*, **84**, 141.
Johansson, S.A., Campbell, J.L. & Malmqvist, K.G. (1995) *Particle-Induced X-Ray Emission Spectrometry (PIXE)*, John Wiley & Sons, New York, Chichester.
Kasahara, M., Ma, C.-J., Kamiya, T., and Sakai, T. (2001) *Nuclear Instruments and Methods in Physics Research B*, **81**, 622.
Khodja, H., Berthoumieux, E., Daudin, L. & Gallien, J-P. (2001) *Nuclear Instruments and Methods in Physics Research B*, **181**, 83–86.
Kobayashi, H. & Gibson, W.M. (1999) *Nuclear Instruments and Methods in Physics Research B*, **152**, 365–369.
McPhail, D.S. (1989) in Vickerman, J.C., Brown, A. & Reed, N.M. Eds., *Secondary Ion Mass Spectroscopy*, Clarendon Press, Oxford, 105–148.
Oakes, A.J.& Vickerman, J.C. (1996) *Surface and Interface Analysis*, **24**, 695.
Steeds, J.W., Gilmore, A., Charles, S., Heard, P., Howarth B. & Butler, J.E. (1999) *Acta Materialia*, **47**, 4025–4030.
Strijckmans, K. (1997) Charged particle activation analysis. In *Surface Characterisation: A Practical Approach*, Hellborg, R., Brune, D., Eds., Scandinavian Scientific Press & VCH, Weinheim, Germany.

Stumborg, M.F., Santiago, F., Chu, T.K., Boulais, K.A. & Price, J.L. (1998) *Nucl. Instr. And Methods in Phys. Res. B*., **134**, 77.

Tadić, T., Jakšić, M., Capiglia, C., Saito, Y & Mustarelli, P. (2000) *Nuclear Instruments and Methods in Physics Research* B, **161–163**, 614–618.

Tadić, T., Jakšić, M., Medunić, Z., Quartarone, E. & Mustarelli, P. (2001) *Nuclear Instruments and Methods in Physics Research* B, **81**, 404–407.

Tirira, J., Serruys, Y., & Trocellier, P. (1996) *Forward Recoil Spectrometry*, Plenum Press, New York, London.

Uzonyi, I., Rajta, I., Bartha, L., Kiss, Á.Z. & Nagy, A. (2001) *Nuclear Instruments and Methods in Physics Research* B, **181**, 193.

Vickerman, J.C., Oakes, A. & Gamble (aka Donsig) H., (2000) *Surface and Interface Analysis* **29**, 349–361.

Watt, F., Grime, G.W., Blower, G.D. & Takacs, J. (1981) *IEEE Transactions on Nuclear Science*, **28**, 1413–1416.

Werner, H.W. (1980) *Surf. Interface Anal.*, **2**, 56.

Yubero, F., Ocaña, M., Caballero, A., & González-Elipe, A.R. (2000) *Acta Materialia* **48**, 4555.

Zink, F., Kerle, T. & Klein, J. (1998) *Macromolecules*, **31**, 417–421.

FURTHER READING

Breese, M.B.H., Jamieson, D.N. & King, P.J.C. (1996) *Materials Analysis Using a Nuclear Microprobe*, John Wiley & Sons, Inc, New York.

Cherepin, V. (1987) *Secondary Ion Mass Spectroscopy of Solid Surfaces*, VNU Science Press, Utrecht.

Grime, G.W. (1999) 'High energy ion beam analysis methods (and background)', in *Encyclopaedia of Spectroscopy and Sprectrometry'* Ed. J.C. Lindon, G.E. Tranter & J.L. Holmes, Academic Press, Chichester, pp. 750–760.

Grime, G.W. (1999) 'Proton microprobe (methods and background)' in *Encyclopaedia of Spectroscopy and Spectrometry*, J.C. Lindon, G.E. Tranter & J.L. Holmes, Ed. Academic Press, Chichester, pp. 1901–1905.

Nastasi, M., Mayer, J.W. & Hirvonen, J.K., (1966) *Ion-Solid Interactions: Fundamentals and Applications*, Cambridge University Press, Cambridge.

Schultz, G. & Weidinger, A. (1996) *Nuclear Condensed Matter Physics: Nuclear Methods and Applications*. Wiley & Sons, Chichester.

Vickerman, J.C., Brown, A. & Reed, N.M. Eds. (1989) *Secondary Ion Mass Spectroscopy*, Clarendon Press, Oxford.

Watt, F., Grime, G.W., Blower, G.D. & Takacs, J. (1981) *IEEE Transactions on Nuclear Science*, **28** 1413–1416.

Wilson, R.G., Stevie, F.A. & Magee, C.W. (1989) *Secondary Ion Mass Spectrometry: A Practical Handbook for Depth Profiling and Bulk Impurity Analysis*, John Wiley & Sons, New York.

Woodruff, R.P. & Delchar, T.A. (1994) *Modern Techniques of Surface Science*, Cambridge University Press, Cambridge.

Chapter 5

Materials Analysis by Electron Beam Probes

Chapter 5
Materials Analysis by Electron Beam Probes

5.1. HIGH RESOLUTION ANALYTICAL ELECTRON MICROSCOPY (HRAEM)

5.1.1 *Interaction of Electrons with Matter*

When a specimen is irradiated with an incident electron beam, it produces a wide range of secondary signals – many of which may be used in AEM to give chemical information about the sample. Figure 5.1 illustrates the situation diagrammatically for the case of a thin specimen, in which some transmission of the incident beam has taken place through the material. The Bremsstrahlung X-rays indicated in Figure 5.1 arise when electrons are decelerated by the Coulomb field of the nucleus of an atom in the specimen, and can possess any energy up to the beam energy, since an electron can suffer any amount of deceleration depending on the strength of its interaction. These X-rays are generally regarded as a useless background signal which only obscures characteristic lines.

The terms *elastic* and *inelastic* scattering of electrons describe that which results in no loss of energy and some measureable loss of energy respectively. If the incident electron beam is coherent (i.e. the electrons are in phase) and of a fixed wavelength, then elastically scattered electrons remain coherent and inelastic electrons are usually incoherent.

The analysis of some of these emitted signals forms the basis of several techniques of local analysis which will be discussed in the following sections. Thus the Auger electrons provide the basis of analysis by Auger Electron Spectroscopy (AES) and the scanning Auger Microprobe (SAM), and analysis of the characteristic X-rays using X-ray energy dispersive spectrometry (XEDS) form the basis of electron microprobe analysis (EMPA) which is another important technique. A modern instrument is essentially an electron optic column which is maintained at high vacuum, with the lenses and most other functions often being controlled by one or more computers.

There are three types of electron microscopes commonly used for microanalysis. These are the scanning electron microscope (SEM) with X-ray detectors, the electron probe microanalyser (EPMA), which is essentially a purpose built analytical microscope of the SEM type, and transmission microscopes (TEM and STEM) fitted with X-ray detectors. In a TEM, compositional information may also be obtained by

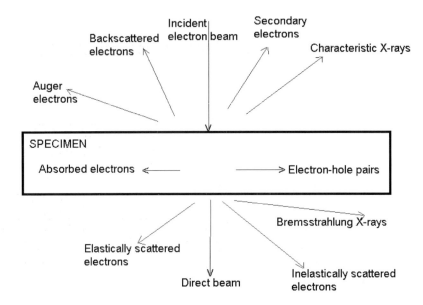

Figure 5.1. Processes occurring when a high-energy beam of electrons interacts with a thin specimen. The arrows do not necessarily represent the physical direction of the signal, but indicate the region in which it may be detected.

measuring the energy loss of the transmitted electrons (EELS) and high resolution electron energy loss spectroscopy (HREELS) will also be discussed.

5.1.2 X-ray Generation

Whenever electrons with several kilovolts of energy strike a solid specimen, 'characteristic' X-rays are emitted, whose wavelengths depend on the nature of the atoms in the specimen, together with white radiation (Bremsstrahlung) of all wavelengths down to a minimum corresponding to the incident electron energy. There is a large number of electron transitions possible in a large atom, each of which leads to the emission of an X-ray of a unique wavelength, as illustrated in Figure 5.2.

Measurement of the wavelength (or energy) of the characteristic X-rays emitted enables qualitative analysis to be carried out. The more difficult process of quantitative analysis requires a measurement of the number of X-ray photons of a given type that are emitted per second.

The $K\alpha_1/\alpha_2$ doublet illustrated in Figure 5.2 is most frequently used for analysis, although as the atomic number of the emitting element increases, the energy required

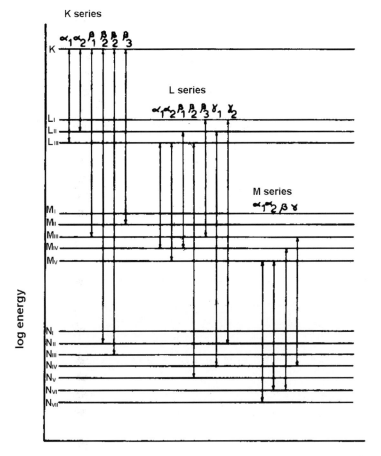

Figure 5.2. Energy level diagram for an atom, showing some of the more common transitions between the K,L,M and N shells which lead to the X-ray lines indicated.

to knock out a K-shell electron also increases. For example, elements heavier than tin ($Z = 50$) are not efficient producers of K X-rays until the incident electron energy is about 75 keV. In an SEM, the electron energies are only of the order 30 keV, so that characteristic L or M X-ray lines would have to be employed in the analysis by this instrument.

5.1.3 Primary Electron Sources

Two kinds of electron sources may be employed in local analysis. The first is a *thermionic* source which produces electrons when heated, and the second is a *field emission* source which produces electrons when an intense electric field is applied to it. Field emission sources give 'monochromatic' electrons, whereas thermionic sources are less monochromatic and may be said to give 'whiter' electrons.

5.1.3.1. Thermionic Sources.

Richardson's Law relates the current density from the source, J, to the operating temperature T:

$$J = AT^2 e^{-\phi/kT}$$

where k is the Boltzmann constant, ϕ the work function and A is Richardson's constant which depends on the source material.

Tungsten has the necessary high melting temperature (3660 K) to be employed as a thermionic source, and lanthanum hexaboride (LaB$_6$) is also employed because of its low work function.

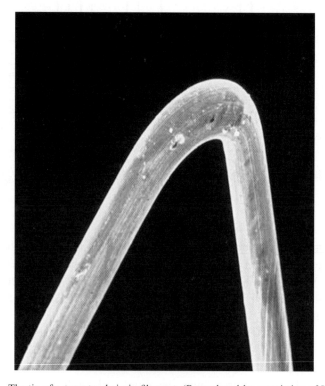

Figure 5.3. The tip of a tungsten hairpin filament. (Reproduced by permission of Wiliams and Carter 1996.)

A tungsten wire filament may be bent into a V shape (a "hairpin" filament, figure 5.3) and its typical lifetime at 100 kV operating voltage is 100 h. LaB$_6$ crystals are grown with a $\langle 110 \rangle$ orientation in order to enhance their electron emission, and their corresponding lifetime is of the order 500 h.

5.1.3.2. Field Emission Electron Sources. The strength of an electric field E is increased at sharp points of radius r, and if a voltage V is applied to such a spherical point, the field at the surface is given by:

$$E = V/r$$

Specimens for field emission sources are of a very fine needle shape, usually in the form of tungsten wire with a tip radius of $< 0.1 \, \mu m$ (Figure 5.4). Application of a potential of 1 kV thus generates a field of 10^6 V/m which lowers the work function barrier sufficiently for electrons to tunnel out of the tungsten. FEG electron microscopes usually employ a gun potential of 3–4 keV.

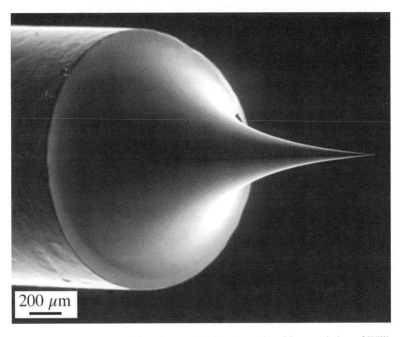

Figure 5.4. A fine tungsten needle forming a FEG tip. (Reproduced by permission of Williams and Carter 1996.)

The tip surface must be free from contaminants, and under UHV conditions the tungsten is operated at ambient temperatures ('cold field emission'). The tungsten can be maintained in pristine condition at a poorer vacuum by heating the tip ('thermal field emission'), when surface treatments with ZrO_2 are applied to improve the emission characteristics. A typical lifetime is at present of the order 10,000 h.

5.1.3.3. Source Brightness. The brightness of an electron source is defined as the current density per unit solid angle. It is a very important parameter when very fine electron beams are used, as in analytical and scanning microscopy. The electrons are focussed after leaving the source, and at this point of crossover one may define the beam diameter d_0, the cathode emission current, i_e, and the semiangle of divergence of the electrons from this point, α_0.

The current density is $i_e / \pi(d_0/2)^2$, the solid angle of the source is $\pi \alpha_0^2$, and the brightness β is defined as:

$$\beta = \frac{4i_e}{(\pi d_0 \alpha_0)^2}$$

The brightness of the three principal sources operating at 100 kV are (in $Am^{-2} sr^{-1}$) 10^9 for tungsten, 5×10^{10} for LaB_6, and 10^{13} for field emission.

5.1.4 Electron Guns

The electron gun controls the beam from the electron source and directs it into the illumination system of the microscope. The design of electron guns is described by Williams and Carter (1996), who discuss in detail their characteristics. For routine TEM applications tungsten sources are excellent – being cheap, robust and easily replaceable. LaB_6 guns are much more expensive, but are to be preferred for AEM because of their increased brightness and longer life.

In cold field emission guns (FEGs), the extremely small source size means that the beam is bright and highly spatially coherent. It is thus the best for AEM, but the necessary UHV conditions make it an expensive technology. Because of the small source size it is not possible in routine TEM to illuminate large areas of the specimen without losing current density and therefore image intensity, so a thermionic source is preferable in this situation.

5.1.5 The Detection and Counting of X-rays
5.1.5.1. Energy-Dispersive Analysis (EDS). EDS is the more generally applicable and versatile system for X-ray analysis, and the detector normally consists of a small

piece of semiconducting silicon or germanium which is held in such a position that as many as possible of the X-rays emitted from the specimen fall upon it.

In an SEM it may be possible to place the detector 20 mm or less from the specimen, but the problems are greater with a TEM because the specimen is within the objective lens. Each incoming X-ray photon excites a number of electrons into the conduction band of the silicon leaving an identical number of positively charged holes in the outer electron shells. The number of electron–hole pairs generated is proportional to the energy of the X-ray photon being detected, and if a voltage is applied across the semiconductor a current will flow as each X-ray is absorbed in the detector. The magnitude of the current will be proportional to the energy of X-ray photon.

The resistivity of the silicon is increased by making the whole detector a semiconductor p-i-n junction which is reverse biased by a potential applied to a thin film of gold on the outer faces. The silicon is doped with a small concentration of lithium, and the whole detector is cooled to liquid nitrogen temperature (77 K). The current which passes between the (gold) electrodes is now very small until an X-ray enters the detector, and the resultant current pulse can be amplified and measured.

The gold-coated outer surface is protected by an ultra thin window of polymer which has been coated with evaporated metal in order to minimise light transmission.

Each amplified pulse is passed to a computer acting as a multichannel analyser (MCA) which registers it in one of ~ 1000 channels, each of which represents a different X-ray energy. The MCA data may then be displayed on a screen in the form of a histogram. The high efficiency of the detector and the relatively large collection angle (~ 0.2 steradians in a TEM) means that data may be collected quite rapidly at quite low beam currents, giving a spectrum within a few minutes. The computer may also store the energies of the X-rays from all the elements, and thus can identify the element giving rise to a line in the spectrum, so that qualitative analysis is quite rapid with such a system.

Quantitative Analysis. The efficiency of the detector is such that almost 100% of the X-rays entering it will produce a pulse, but the *pulse processing speed* limits the rate at which X-rays can be counted. If the count rate is less than a few thousand counts per second, then most of the incoming pulses are processed, but as the count rate rises an increasing fraction of the pulses are rejected. The *live time* during an analysis when the detector was counting is thus less than the *elapsed time*, and the EDS system records both times in order that the true count rate may be measured.

The energy resolution of the detector is relatively poor, so that each X-ray line appears as a broad peak 100–200 eV wide. Any X-ray line thus occupies several channels of the MCA, and so the peak height is reduced. This factor, together with a low peak-to-background ratio limits the use of the EDS system for quantitative analysis.

5.1.6 *Wavelength Dispersive Spectrometer (WDS)*

In a WDS, the X-radiation coming from the specimen is filtered by means of a curved crystal spectrometer, which employs diffraction to separate the X-rays according to their wavelength. A typical arrangement for the spectrometer is shown in Figure 5.5.

The X-rays leave the specimen at a take-off angle ϕ, are collimated by two slits S_1 and S_2 before falling on to a crystal (bent to a radius $2R$, where R is the

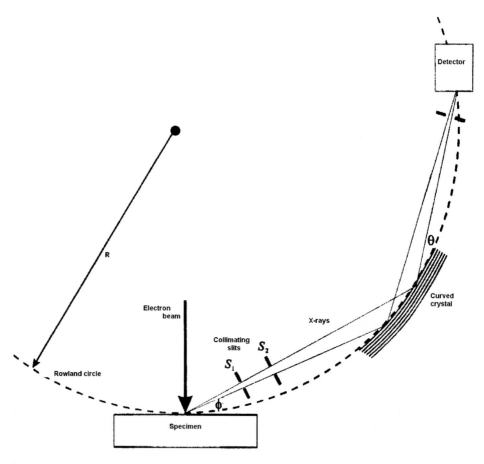

Figure 5.5.   A wavelength-dispersive spectrometer showing the specimen, crystal and detector all lying on the Rowland circle of radius R.

radius of the Rowland circle illustrated in Figure 5.5) of lattice spacing *d*. If the angle between the incident X-rays and the crystal lattice planes is θ, then the only X-rays diffracted to the detector will be of wavelength given by the Bragg equation:

$$\lambda = \frac{2d \sin \theta}{n}$$

where *n* is the order of the diffracted beam.

The spectrometer is set to the appropriate Bragg angle θ of the requisite characteristic wavelength, and only these X-rays will reach the detector and be counted. The detector employed is the gas proportional counter, which can operate at much faster count rates than the EDS crystal detector.

The spectrometer is necessarily quite large, and a complicated mechanism has to be precision engineered in order to enable θ to be altered while keeping both the crystal and the detector on the Rowland circle. In order to cover the whole X-ray spectrum a range of crystals with different lattice spacings is required, which may be interchanged automatically.

WDSs have excellent resolving power, and the peak-to-background ratio of each line is much higher than can be achieved with a crystal detector. With a suitable crystal of large lattice spacing it is possible to detect and count X-rays as soft as boron K_α or even beryllium K_α, and this type of spectrometer is widely used when accurate quantitative analysis is desired.

In the following sections we will consider in turn the instruments employed for electron beam microanalysis.

5.2. ELECTRON PROBE MICROANALYSIS IN THE SCANNING ELECTRON MICROSCOPE

5.2.1 *Instrumentation*

Microprobe analysis was initially developed at the University of Paris by R. Castaing, who fitted an X-ray spectrometer to a converted electron microscope in the early 1950s, and the first commercial instrument, developed in France by the Cameca company, appeared in 1958. The following years saw commercial instruments produced in the UK, USA and Japan.

A typical instrument is equipped with four computer controlled crystal spectrometers, as well as an EDS system for preliminary qualitative analysis. A light microscope is provided for examining the specimen and also for ensuring that the specimen height is adjusted until it is on the Rowland circle. The drawing of

Figure 5.6 illustrates the arrangement of the components in a modern commercial microanalyser.

The usual source of electrons is a tungsten filament electron gun held at a negative potential (typically 10–30 kV), and magnetic lenses focus the beam into a fine probe incident on the surface of the specimen. A probe diameter of 0.2–1 μm is typical, with a current of 1–100 nA.

The specimen stage usually holds several specimens and standards, with dimensions typically 20–30 mm, although for special purposes extra-large specimen stages are available – Matsuya *et al.* (1988) describe a holder with a working area of 300 × 300 mm. An airlock isolated from the main vacuum chamber reduces the time taken to change specimens.

The specimen must have a conducting surface, so that non-conducting specimens must therefore be coated before being placed in the instrument. Since the coating will itself absorb X-rays as they are emitted from the specimen, and it will also emit its own characteristic X-rays, it should be as thin as possible and be of low atomic

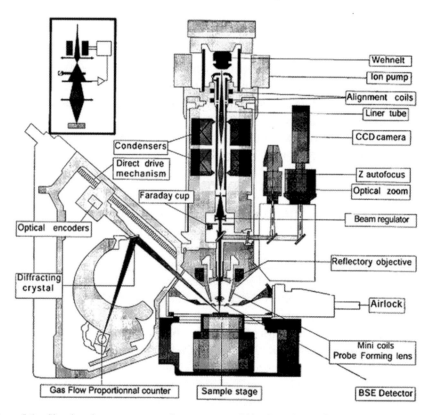

Figure 5.6. Showing the arrangement of components within the column of a Cameca SX100 EPMA.

weight. A thin coating of carbon is therefore usually applied to non-conducting specimens by vapour deposition in a vacuum.

Flat polished specimens are necessary, as X-rays will not be detected from regions of the specimen which are not in line-of-sight of the detector. Rough specimens would therefore give topographical effects which might be misinterpreted as variations in composition.

Orthogonal x and y movements of the specimen are required, and computer control of the specimen position under the beam enables large numbers of points to be analysed without intervention of the operator.

EPMA instruments are equipped with a number of microscopy tools that allow simultaneous imaging of the characteristic X-rays (by both wavelength and energy dispersive analysis), and also of the secondary electrons as well as the back-scattered electrons (BSE). These, together with the sophisticated visible light optics provide very flexible sample inspection with image magnification ranging from 40× to 400,000×.

5.2.2   Point or Spot Analysis

The spatial resolution in quantitative analysis is defined by how large a particle must be to obtain the required analytical accuracy, and this depends upon the spatial distribution of X-ray production in the analysed region. The volume under the incident electron beam which emits characteristic X-rays for analysis is known as the *interaction volume*. The shape of the interaction volume depends on the energy of the incident electrons and the atomic number of the specimen, it is roughly spherical, as shown in Figure 5.7, with the lateral spread of the electron beam increasing with the depth of penetration.

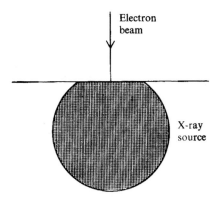

Figure 5.7.   Illustrating the shape of the volume of specimen under the incident electron beam from which characteristic X-rays are emitted.

Reed (1966) has derived the following general expression for resolution d (in μm):

$$d = 0.077(E_0^{1.5} - E_c^{1.5})/\rho$$

where E_0 is the initial energy of the electrons in keV, E_c (in keV) is the critical excitation energy of the characteristic X-rays in question and ρ is the density (in g cm^{-3}). It can then be shown that the spatial resolution for quantitative analysis (defined as the particle size required to contain 99% of the X-ray production) is about three times d as calculated from the above equation.

Reed (1966) produced the nomogram of Figure 5.8 for spatial resolution (d) as a function of density, incident electron energy and critical excitation energy. In the diagram, the case for iron ($E_c = 7.1$ keV, $\rho = 7$) gives an estimated value of d at 20 keV of 0.8 μm, so the resolution for quantitative analysis is approximately 2.4 μm.

This point analysis is employed to analyse a selected region of chemically homogeneous composition, such as a phase. The electron beam is stopped and positioned carefully on the point selected on the SEM screen, and the composition of the sampling volume is determined by a crystal spectrometer.

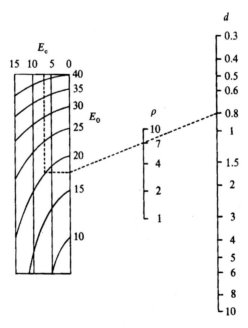

Figure 5.8. Nomogram for spatial resolution (d), in μm, as a function of density (ρ), in g/cm^3, incident electron energy (E_0) and critical excitation energy (E_c), both in keV. The case for 20 keV electrons in iron is illustrated. (Reed, 1966).

5.2.3 Scanning Analysis

The probe may be scanned over the specimen surface by means of electromagnetic coils driven by a waveform generator, which also supplies a synchronous signal to the display. By modulating the brightness of a cathode ray tube with the X-ray spectrometer, a scanning image showing the spatial distribution of a selected element may be produced.

For an EDS system the beam may be scanned over, say, $100 \times 100 \, \mu m$, and the analysis is then an average of the area of the image on the screen. For a WDS system such scanning would excite part of the specimen surface not lying on the Rowland circle (Figure 5.5), so if an analysis of an area greater than about $5 \times 5 \, \mu m$ is required by WDS, then the specimen must be scanned, and not the beam.

5.2.3.1. Line Scanning.

In order to examine the variation of chemical composition within a sample, one approach is to select the X-ray signal from the element of interest and to display its intensity as the point of incidence of the electron beam is moved along a chosen path (by moving either the beam or the specimen). The instantaneous count rate is measured, and a trace of composition versus beam position is obtained, as illustrated in Figure 5.9.

This EPMA line scan was analysed by wavelength dispersive spectroscopy, being part of a study by Hörz and Kallfass of ornamental and ceremonial artifacts dated to approximately AD 50–300, recovered from the Royal Tombs of Sipán, Peru.

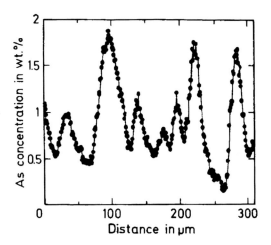

Figure 5.9. Examination of segregation bands of arsenic in copper in archaeological artefacts
(Hörz and Kallfass, 2000).

Copper pellets were found inside a series of objects such as beads, and etched metallographic sections were prepared. The specimen whose analysis appears in Figure 5.9 showed a recrystallized structure showing zones of segregation worked out into bands. The line scan shows that the segregation is reflected by periodical changes in the arsenic content – the concentrations ranging between 0.2% and 1.8%. These structural features give evidence that the cast copper pellet was brought into its final shape by working and was subsequently annealed.

An accurate analysis is difficult by this technique, because the electron beam spends only a short time on each spot, and so the counting statistics will not be adequate for data of high precision to be accumulated.

5.2.4   X-ray Mapping

It is possible to extend the line scanning method to two-dimensional scanning. In its simplest form, the display is made bright every time an X-ray photon is counted, thus generating a image of bright dots. The dot density provides a qualitative measure of the concentration of the element of interest.

5.2.4.1. Digital X-ray Mapping.

The beam may be controlled by a computer to move within a grid of points on the specimen surface. The beam remains at each point for a pre-set time while an analysis is carried out. With an EDS system, data from several elements may be collected at the same time and a BSE image may also be acquired in the same experiment. Digital X-ray maps may then be displayed for each of the chosen elements in turn.

5.2.4.2. Colour Images.

The present generation of software available with WDS in the EPMA has many options in digital mapping. Figure 5.10 shows stage-scanned maps of a cross-section through an archaeological specimen in the form of a cross-section through the blade of a Roman mattock.

The silicon map shows the distribution of slag inclusions, while the position of the weld lines can be seen to be marked by boundaries between areas of high and low phosphorus content and by enrichments of nickel and/or arsenic at the weld lines. Higher resolution maps can be made of welds to reveal their fine structure.

An example of a higher resolution map from the EPMA is given in Figure 5.11, where the sample is again an archaeological specimen – a pair of shears from the same site as the specimen shown in Figure 5.10. Here the weld lines are marked by lines of sulphide and slag particles in an area depleted of phosphorus.

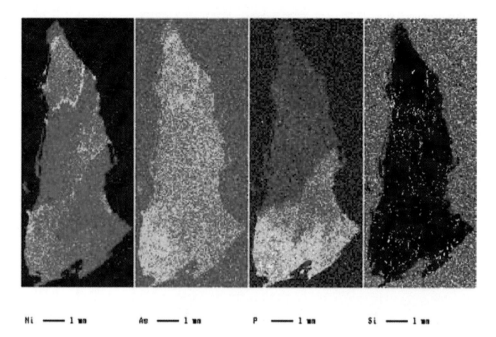

Ni ——— 1 μm   As ——— 1 μm   P ——— 1 μm   Si ——— 1 μm

Figure 5.10. EPMA maps for nickel, arsenic, phosphorus and silicon in a 1st century AD Roman mattock blade. Scale bar = 1 mm. (Courtesy Dr C.J. Salter, Oxford University.)

Figure 5.11 illustrates another use of colour is the representation of intensities by dividing the intensity scale into bands and assigning a colour to each, thus enhancing the visibility of intensity variations across the area of specimen under examination.

5.2.5 Corrections in Quantitative Analysis

The uncorrected concentration (C′) of each element is given by:

$$C' = C_0(I/I_0)$$

Where I and I_0 are the characteristic X-ray intensities measured from the unknown specimen and standard respectively, and I/I_0 is commonly known as the 'k ratio'. C_0 is the concentration of the element concerned in the standard.

The composition of the analysed point is calculated from the *corrected* intensities by applying 'matrix corrections' which take account of a number of factors

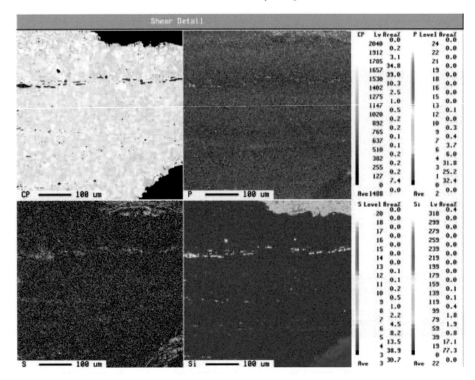

Figure 5.11.   EPMA maps of Roman shears (CP = atomic number contrast). (Courtesy Dr C. J. Salter, Oxford University.)

governing the relationship between intensity and composition. These may be represented by three factors, F_a, F_f, F_b and F_s:

F_a arises from absorption of characteristic X-rays emerging from the specimen,

F_f allows for an enhancement of the characteristic X-ray intensity due to fluorescence,

F_b accounts for a loss of X-ray intensity arising from some of the incident electrons being backscattered out of the specimen, and

F_s is the 'stopping power' factor. Electron penetration is a function not only of the incident electron energy (which is constant for a given analysis) but also on the 'stopping power' of the sample, which depends somewhat on atomic number. Reed (1993) derives equations for the generated characteristic X-ray intensity, leading to expressions for F_s.

5.2.5.1. Absorption Corrections.   If I' is the intensity of a collimated X-ray beam of initial intensity I, after passing through a thickness x of material of density ρ, the power of a material to absorb X-rays is expressed in terms of the *mass attenuation coefficient* μ in the expression:

$$I' = I \exp(-\mu\rho x)$$

In EPMA, X-rays are produced over a range of depths from the surface before their energy falls below E_c. In order to derive the effective absorption factor, an integration must be carried out which requires a knowledge of the shape of the depth distribution of X-ray production.

The depth distribution function, $\phi(\rho x)$ represents the X-ray intensity per unit mass depth (ρz), relative to that in an isolated thin layer, and is of the form illustrated in Figure 5.12.

If ψ is the X-ray takeoff angle in the instrument, then the absorption factor $f(\chi)$, where $\chi = \mu \operatorname{cosec}\psi$, may be written (Reed 1993):

$$f(\chi) = \int_0^\infty \phi(\rho z) \exp(-\chi\rho z) \, d(\rho z) \Big/ \int_0^\infty \phi(\rho z) \, d(\rho z)$$

The denominator in this equation represents the area under the curve of Figure 5.12 and serves to normalise the expression, so that $f(\chi) \to 1$ as $\chi \to 0$. The absorption correction factor F_a is equal to $1/f(\chi)$.

5.2.5.2. Fluorescence Corrections. Fluorescence occurs when the characteristic radiation of a given element (A) is excited by X-ray photons of higher energy than

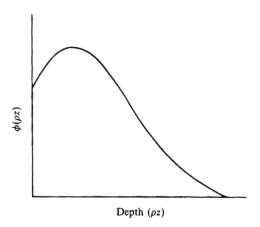

Figure 5.12. The X-ray depth distribution function $\phi(z)$.

the critical excitation energy (E_c) of A. The most significant source of such photons is the characteristic radiation of other elements present in the sample. For example, for given elements of atomic number $Z_A \leq 21$, fluorescent K-shell ionisation can be caused by the K radiation of any element of higher atomic number, i.e. the condition for excitation is $Z_B - Z_A > 1$, where Z_B is the atomic number of the exciting element. The same principles apply for excitation by L lines, and although fluorescence involving M radiation also occurs, in practice its contribution is small and can be neglected.

In order to apply fluorescence corrections it is necessary to calculate the intensity of the fluorescence radiation of element A (I_f) excited by the characteristic radiation of element B, and to obtain the ratio I_f/I_A, where I_A is the intensity of 'A' radiation produced directly by electron bombardment. The fluorescence factor F_f is given by $F_f = 1/[1 + (I_f/I_A)]$.

5.2.5.3. Backscattering Corrections.

The backscattering correction factor F_b is given by:

$$F_b = 1/(1 - \eta_x) = 1/R$$

where η_x is the fractional X-ray intensity loss, i.e. the intensity lost divided by the total intensity that would be obtained with no backscattering. R $(= 1 - \eta_x)$ is the factor by which the X-ray intensity is reduced owing to electron backscattering.

Springer (1976) proposed the following polynomial for R:

$$R = a_0 + a_1 W_0 + a_2 W_0^2 + a_3 W_0^3 + a_4 W_0^4$$

With coefficients calculated from the expression:

$$a_i = b_{0i} + b_{1j}Z + b_{2i}Z^2 + b_{3i}Z^3 + b_{4i}Z^4$$

Figure 5.13 shows R as a function of Z for different values of W_0 $(= E_c/E_0)$ obtained from this polynomial.

The overall correction factor is given by the product of these individual factors:

$$F = F_a F_f F_b F_s$$

So the true concentration (C) in the specimen may be written:

$$C = C'(F/F_0)$$

where F_0 is the overall matrix correction factor for the standard.

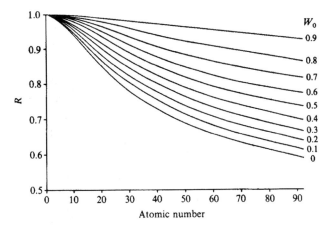

Figure 5.13. Backscattering correction factor R as a function of atomic number and W_0 $(= E_c/E_0)$
(Springer 1976).

5.2.5.4. Iteration. Matrix correction factors are dependent on the composition of
the specimen, which is not known initially. Estimated concentrations are initially
used in the correction factor calculations and, having applied the corrections thus
obtained the calculations are repeated until convergence is obtained, i.e. when the
concentrations do not change significantly between successive calculations.

Current commercial systems normally provide proprietary computer software
which general incorporates the correction methods outlined above.

5.3. ANALYTICAL TEM

5.3.1 X-ray Analysis in the TEM

The spatial resolution of X-ray analysis carried out in the EPMA is limited to the size
of the sampling volume, which is around $1\,\mu m^3$. There may be many important
features of a specimen which are smaller than $1\,\mu m$, and one way of overcoming the
problem is by the use of thin specimens. We have seen (Figure 5.7) that the lateral
spread of the electron beam increases with depth of penetration, so that in a
sufficiently thin specimen the beam spread is much less. We will therefore next
consider the analysis of thin foil specimens in the TEM.

If the sampling volume is now treated as a truncated cone, and if the specimen
thickness is t nm, then the lateral beam spread B (in nm) for thin specimens is given by:

$$B = 0.198(Z/E)(\rho/A)^{1/2}t^{3/2}$$

Where $Z =$ atomic number, $A =$ atomic weight, E the electron energy (keV), and ρ
the density (g/cm^3).

With a thermionic electron source, and a foil thickness of 100 nm, the volume of specimen excited is of the order $10^{-5}\,\mu m^3$. With a FEG source in a dedicated STEM, however, with a foil of 1 nm thickness this specimen–beam interaction volume can be as small as $10^{-8}\,\mu m^3$. Very small signal levels are thus to be expected in AEM, hence the importance of employing higher brightness sources and the need to modify the specimen–detector configuration to maximize the collection angle.

5.3.2   Specimen Preparation for TEM

The most comprehensive textbook on TEM is that by Williams and Carter (1996). It is written for the microscope operator, and, as well as explaining the use of the instrument for imaging, diffraction and spectroscopy, it also gives an account of the principles of specimen production, and the reader is encouraged to turn to that book for a fuller account of the subject.

The final TEM specimen must of course be electron transparent, and it must be either self-supporting or it must rest upon some form of support grid (usually made of copper). The specimen or grid will usually be 3 mm in diameter.

Self-supporting disks are prepared in several stages: an initial slice of material is prepared between 100 μm and 200 μm in thickness. Ductile materials may be rolled, cut with a chemical saw, or spark eroded to produce a thin slice; brittle materials may be cleaved with a razor blade or cut with an ultramicrotome (see below). The 3 mm disk is then cut from the slice, using a mechanical punch for ductile materials, while brittle materials need to be spark eroded or drilled.

The disk is then prethinned at its centre ('dimpling') to produce a region ∼ 10 μm thick. This may be done either mechanically or chemically. The final stage of thinning of the disk is aimed at producing an electron-transparent area at the base of the dimple. Electrically conducting samples may be finally thinned by electrolytic polishing, although it must be borne in mind that this may change the surface chemistry of the specimen. Another important and versatile technique is that of ion milling, in which the prethinned specimen is bombarded with energetic ions in order to sputter material away until it is thin enough to be studied in the TEM. Fibres and powders may also be thinned by ion milling if they are first embedded in epoxy resin, held in a 3 mm brass tube and then cut into 3 mm disks which are then ion milled to electron transparency.

Specimens on grids are an alternative to self-supporting disks. Specimens in the form of small particles may be supported on a thin film (e.g. an amorphous carbon film) before being placed on the grid. Thin slices of the material may be produced by an instrument known as an *ultramicrotome*. The principal advantage of the technique is that it leaves the chemistry unchanged, and it may be employed to create uniform thin films of multiphase material. The instrument operates by moving the specimen past a knife blade. The blade can be of glass for soft materials, but will be diamond for harder ones. The thin flakes float off into water or other inert medium from where they are collected on grids.

In a 100 kV TEM, useful specimens can range in thickness from ~20 nm to 0.5 μm. For microanalysis, the thinner the specimen the better the quality of the information that can generally be obtained.

The need to be able to thin complex microelectronic devices, and to select and thin specific regions within them has resulted in ever-more sophisticated specimen preparation methods involving precision ion polishing. This requirement culminated in the development of the focused ion beam (FIB) technique, which is able to slice out electron-transparent foils from any multilayer, multiphase material with extreme precision. Overwijk *et al.* (1993) have described such a technique for producing cross-section TEM specimens from (e.g.) integrated circuits.

They employed a FIB of 30 keV Ga ions which as focused to a spot with a diameter which could be varied between ~0.05 and ~1 μm. The beam current varied with focus size between 13 pA and 1.2 nA, and sputter rate typically increased with beam current from 0.005 to 0.5 μm^{3}s^{-1}. The beam control was automated, so that the major part of the specimen preparation was performed automatically, only the final high-resolution operations being carried out by manual adjustment of the milling area. Their preparation scheme was as follows:

Having selected an area of interest, a staircase-shaped recession was sputtered on both sides of the future TEM specimen (Figure 5.14).

The staircases were constructed by sputtering a sequence of rectangular recesses of decreasing width, aligned with the side closest to the TEM specimen. The dimensions of the staircase were typically 10×6 μm^2 for the top rectangle, 10×2 μm^2 for the bottom one, and a total depth of several microns. The TEM specimen was subsequently reduced in width with a high-resolution beam to a thickness between $0.2 - 0.3$ μm, and finally disconnected from the substrate by tilting the specimen stage by 45° of arc, and 'cutting' with a focussed FIB around the edges to detach the membrane from the bulk material. This procedure resulted in a TEM specimen within 2 h. In a review, Newbury and Williams (2000) point out that this tool costs as much as a basic TEM but, for the study of complex materials it is now as essential as the microscope itself.

5.3.3 The Conventional Transmission Electron Microscope

Figure 5.15 shows a ray diagram for a light-optical projection microscope. The light source is placed behind a condenser system which collects the light which is diverging from the source and illuminates the specimen. The presence of the variable aperture near to the condenser lens permits control of the area of the specimen which is

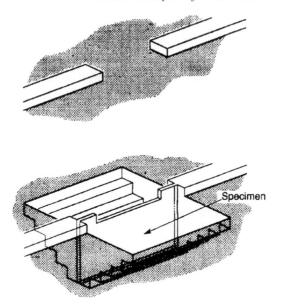

Figure 5.14. TEM specimen preparation scheme by FIM milling, illustrated for a specimen from an integrated circuit. The specimen is formed by sputtering two staircase-shaped recessions on either side. The dimensions of the specimen are typically $10 \times 5 \times 0.2\,\mu m^3$. (Reproduced by permission of Overwijk *et al.* 1993.)

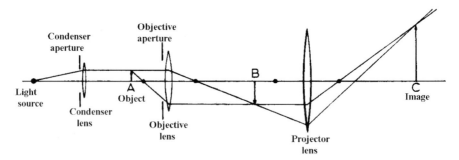

Figure 5.15. The optical system for a transmission projection microscope.

illuminated, and the variable aperture near the objective lens controls the angular spread of light which is collected from the specimen.

The electron optics in TEM is the same as the light optics of Figure 5.15, with the obvious difference that the wavelengths differ by many orders of magnitude: visible light is of wavelength 400–700 nm, while electrons have wavelengths in the range 0.001–0.01 nm. The resolution limit (δ, the smallest observable distance between two points in an image) of the TEM is thus theoretically several orders of magnitude smaller than that of the light microscope. Because electrons carry a charge, electromagnetic fields can be used as lenses for electrons, and since electrons are strongly scattered by gases, all the optical paths must be evacuated to a pressure $< 10^{-6}$ Pa. In practice, electron lens defects limit the resolution of the TEM to ~ 0.1 nm in the highest-resolution instruments.

Figure 5.16 shows a cross-sectional view of a typical 200 keV TEM. By means of condenser lenses, the illumination system can produce a demagnified image of the source in the specimen plane. Normal TEM imaging uses an out-of-focus source image, the image plane of which is well away from the specimen plane. By means of limiting aperture diaphragms within the lens, an approximately parallel, coherent electron beam is produced at the specimen.

The analytical electron microscope (AEM) is fitted with a spectrometer for X-ray microanalysis and also for electron energy-loss analysis (q.v. pp. 185).

5.3.4   The Scanning Transmission Electron Microscope (STEM)

Most modern AEMs are based on a STEM, which scans a fine probe of electrons across the specimen. The so-called dedicated STEM (DSTEM) only operates in scanning mode, whereas modified TEM (TEM/STEM) instruments are more numerous. Both types of instrument employ a FEG source.

Figure 5.17 illustrates the principle of the STEM imaging system: the probe remains parallel to the optic axis of the microscope as it scans. Two pairs of scan coils are used to pivot the beam about the front focal plane of the objective polepiece.

The basic principle of image formation in the scanning mode is to employ connected scanning coils to scan a CRT synchronously. The STEM signal generated at any point on the specimen is detected and amplified. It is usual to make use of the transmitted and the scattered electrons to form images; a proportional signal is then displayed at an equivalent point on the CRT screen. This is the same principle as that used in the SEM (q.v.). Having created a STEM image in this way, the scanning probe can be stopped and positioned upon the feature to be analyzed.

5.3.4.1. X-ray EDS in the STEM.

There is insufficient room in the stage of the STEM to introduce a wavelength dispersive spectrometer, with its very high energy resolution. The X-rays emitted by the excited atoms are therefore detected using an energy-dispersive spectrometer (EDS) consisting of a liquid nitrogen-cooled crystal of silicon doped with lithium. Elements with an atomic number $Z < 4$ cannot be

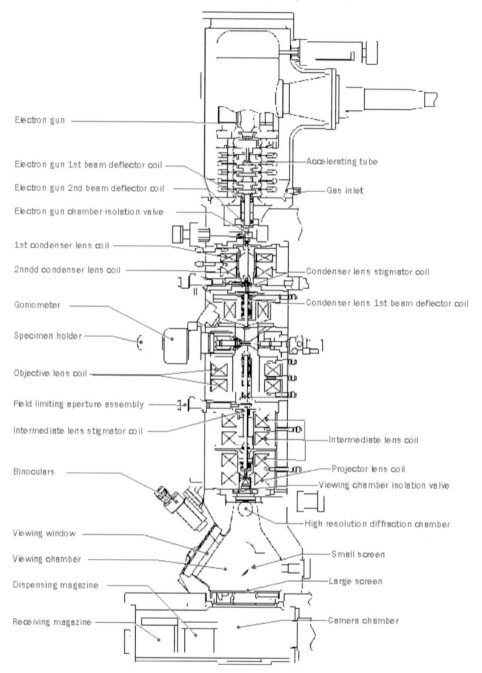

Figure 5.16.   Cross-section of the column of a 200 kV FEG TEM (Courtesey of JEOL).

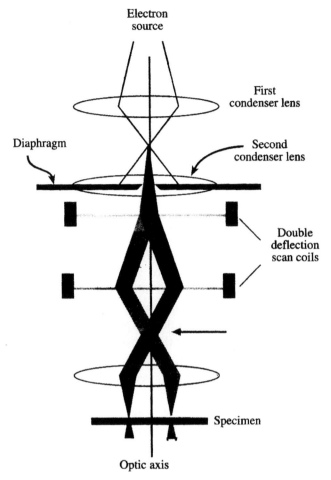

Figure 5.17. Illustrating the principle of STEM image formation using two pairs of scan coils between the second condenser lens and the upper objective polepiece. (Reproduced by permission of Williams and Carter 1996.)

detected, and quantification of the signal from elements with $Z < 8$ is difficult because of the inefficiencies of both the X-ray generation and detection process.

An EPMA is able to provide a quantitative analysis with an accuracy of ± 1–2%, but the AEM is usually relatively poorer, with an accuracy which depends very much on the particular element and its concentration. However, the actual mass analysed by AEM can be as small as 10^{-22} g.

Ideally the EDS should only receive the X-rays from beam–specimen interaction volume, but it is not possible to prevent radiation from the microscope stage and from other areas of the specimen from entering the detector. The analysis of bulk specimens in the EPMA approaches this ideal, but the higher accelerating voltage employed in the AEM generates intense stray Bremsstrahlung X-rays in the illuminating system and also uncollimated electrons. These may strike the specimen outside the area of interest within the interaction volume and produce significant amounts of spurious X-rays. If the focussed electron beam is positioned down in a *hole* in the specimen, such stray radiation would give rise to an X-ray spectrum characteristic of the specimen. This is a standard method of assessing the 'cleanliness' of the illumination system of the AEM, and the hole count (intensity) should always be < 1% of that of the specimen.

5.3.5 *Qualitative X-ray Analysis of Thin Specimens*

Having obtained a spectrum across the energy range all the characteristics peaks of the specimen, the individual peaks are identified. The computer system of the AEM may often provide an automatic identification display, if the system is well calibrated. If the spectrum contains many peaks, it may be necessary to follow a manual sequence of peak identification which was developed by Goldstein *et al.* (1992). This involves looking at the most intense peak and then working down through its family (e.g. if a K_α matches the peak, then the K_β sought, followed by the L lines etc.), this being followed by the next most intense peak, and so forth.

Li *et al.* (2000) have employed nanometer scale analysis in a FEG-TEM operating at 200 kV to distinguish between true GP zones in an Al-Zn-Mg-Cu alloy and 'GP zone-like' defects caused by electron beam irradiation in the TEM. They studied an Al-6.58Zn-2.33Mg-2.40Cu (wt%) alloy, in which it is well known that the decomposition of supersaturated solid solutions takes place via the formation of GP zones, using conventional techniques to produce thin foil specimens of aged material.

Figure 5.18(a) is a high-resolution image of a GP zone in this alloy, lying on the {111} plane, and Figure 5.18(b) shows a 'GP zone-like defect' caused by electron irradiation. It is difficult to distinguish the GP zone and the defect structure using conventional AEM because they are too small to identify. Li *et al.* (2000) employed a beam spot size of 0.5 nm to perform an EDXS analysis, and Figure 5.19 shows the spectra obtained.

Comparison of the two spectra confirms that there is an enrichment of Zn and Mg in the GP zone in the upper illustration. There is, however, no sign of Zn or Mg enrichment in the 'GP zone-like' defect structure, as shown in the lower spectrum.

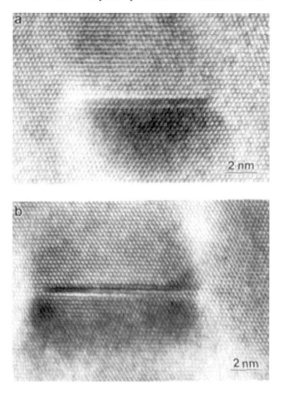

Figure 5.18. High resolution TEM images of (a) a GP zone and (b) a GP zone-like defect in aged AlZnMgCu alloy (Li *et al.*, 2000).

5.3.6 *Quantitative Analysis of Thin Specimens (Cliff-Lorimer and ζ Methods)*

All quantitative X-ray microanalysis requires that the background Bremsstrahlung be subtracted from the observed spectrum, so that the relative intensities of the characteristic X-rays can be determined.

5.3.6.1. *Background Subtraction.*

If the characteristic peaks to be measured appear isolated in the spectrum, so that there are regions in the background which are clear of characteristic peaks, an empirical background can be fitted by computer on to the spectrum when it is displayed on the VDU. This method gives better results with greater intensity levels in the spectrum.

In complex materials, where peaks are situated very close to each other, or when the peaks are below $\sim 1.5\,\text{keV}$, this method is not practicable, and a mathematical modelling approach is made. Modelling the Bremsstrahlung intensity produces a

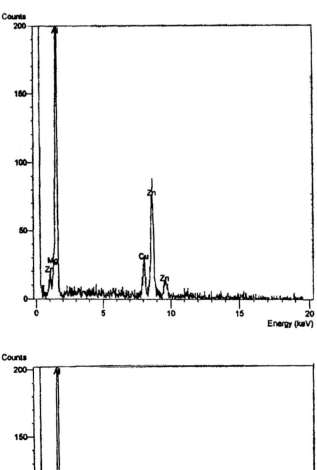

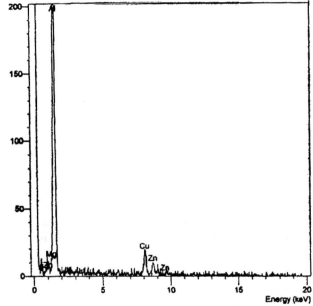

Figure 5.19. EDXS spectra corresponding to the images in Figure 5.18(a) and (b) respectively.

smooth curve fit, above which the intensity of the characteristic lines can be estimated.

5.3.6.2. The Thin Film Approximation.

If absorption and X-ray fluorescence are neglected (i.e. the film is assumed to be 'infinitely thin'), then the measured X-ray intensity from element A in a thin specimen of unit thickness, generated by a current i is given by

$$I_A = n_A Q_A \omega_A a_A \eta_A i$$

And for element B by

$$I_B = n_B Q_B \omega_B a_B \eta_B i$$

where Q is the ionisation cross-section (i.e. the probability of ejecting the electron) and is the fluorescent yield (i.e. the probability of the ionised atom returning to the ground state by emitting the specific X-ray photon). a and η represent the fraction of the K (or L and M) line which is collected and the detector efficiencies respectively for elements A and B

Thus in an alloy made up of elements A and B

$$\frac{n_A}{n_B} = \frac{I_A Q_B \omega_B a_B \eta_B}{I_B Q_A \omega_A a_A \eta_A} \qquad (1)$$

which is generally written as

$$\frac{n_A}{n_B} = k_{AB} \frac{I_A}{I_B}$$

Cliff and Lorimer (1975) used this equation to form the basis for X-ray microanalysis of thin foils, where the constant k_{AB} contains all the factors needed to correct for atomic number differences. k_{AB} varies with operating voltage, but is independent of sample thickness and composition if the two intensities are measured simultaneously. Its value can be determined experimentally with accuracy, using specimens of known composition. The value of k_{AB} can be determined by calculation more rapidly, but with less accuracy.

On the assumption that the background bremsstrahlung has been removed, and that the data are in the form of the measured intensities at full width half maximum (FWHF), equation (1) can be used to calculate the ratio of the number of atoms of A to the number of atoms of B, i.e. the concentrations in the binary material.

Standard thin films of known composition may not always be available, and in multi-component systems many k factors should be determined, which is a time-consuming process. A new quantitative procedure for thin specimens has been developed to overcome these limitations:

In a thin specimen, the magnitude of I_A is proportional to the mass thickness, ρt:

$$\rho t = \zeta_A (I_A / C_A)$$

where C_A is the weight fraction of A, and ζ_A (zeta) is the factor connecting I_A to ρt. With respect to element B we can write:

$$\rho t = \zeta_B (I_B / C_B)$$

where C_B is the weight fraction of B. When ζ factors for A and B are known, and assuming that $C_A + C_B = 1$, we can write:

$$C_A = \frac{\zeta_A I_A}{\zeta_A + \zeta_B I_B}, \qquad C_B = \frac{\zeta_B I_B}{\zeta_A I_A + \zeta_B I_B}, \qquad \rho t = \zeta_A I_A + \zeta_B I_B$$

Therefore, C_A, C_B and ρt can be determined simultaneously by measuring X-ray intensities (if the specimen density and thickness are known). Only ζ factors are required and k factors are not used. This approach can be extended to any multi-component system if one assumes $\sum C_I = 1$. The ζ factors are measured from standard thin films with known composition and thickness, the advantage being that pure element thin films can be applied as standards.

The Cliff–Lorimer and the ζ methods break down when the thin foil assumptions are invalid. Absorption and fluorescence corrections may have to be made if thicker foils are employed.

5.3.6.3. *Absorption Correction.*

The absorption correction factor (ACF) is the ratio of the effective sensitivity factor $k_{AB}*$ to the true sensitivity factor k_{AB}, and is given by:

$$\text{ACF} = \frac{\int_0^t \left\{ \varphi_B(\rho t) e^{-(\mu/\rho)_{Spec}^B \rho t \, cosec\alpha} \right\} d(\rho t)}{\int_0^t \left\{ \varphi_A(\rho t) e^{-(\mu/\rho)_{Spec}^A \rho t \, cosec\alpha} \right\} d(\rho t)}$$

In this expression, $\mu/\rho]_{Spec}^A$ is the mass absorption coefficient of X-rays from element A in the specimen, α is the detector take-off angle, ρ is the density of the specimen and t is the thickness. The term $\phi(\rho t)$ is the depth distribution of X-ray production, which is the ratio of the X-ray emission from a layer of element A of thickness $\Delta\rho t$ at a depth t in the specimen to the X-ray emission from an identical, but isolated, film. For most thin foils $\phi(\rho t)$ may be assumed to be constant and equal to unity, that is, a uniform distribution of X-rays is generated at all depths throughout the foil.

Because the sample density and the values of μ/ρ vary with the composition of the specimen, the complete absorption correction procedure is an iterative process.

5.3.6.4. Fluorescence Correction. Fluorescence is usually a minor effect, and is often ignored. The classic case is that of Cr in stainless steels, where the Cr K_α line us fluoresced by the Fe K_α line, giving rise to an apparent increase in Cr content as the foil gets thicker.

5.3.6.5. Composition Profile Measurement. Results of Zieba et al. (1997) will be given as an example of the measurement of solute distribution in an alloy undergoing a phase transformation. They studied discontinuous precipitation in cobalt–tungsten alloys, in which a Co-32 wt% W alloy was aged in the temperature range 875 K to 1025 K, and high spatial resolution X-ray microanalysis of thin foils by STEM was used to measure the solute distribution near the reaction front.

The cellular reaction product consisted of alternating plates of Co_3W lamellae in a solute-depleted Co matrix (ε_{Co}). Homogenized ingots were cut into slices 1 mm thick and then heat-treated. Spark erosion was used to trepan 3 mm diameter discs which were then jet electropolished to form thin foils for TEM examination in a Philips EM 430 STEM instrument operating at 300 kV in the nanoprobe mode with a probe size of 5 to 10 nm.

Energy-dispersive X-ray analyses were carried out by using the ratio technique (equation 1) to relate compositions to the intensities of the CoKα and WLα characteristic X-ray peaks. The value of the k_{W-Co} factor was experimentally determined using single-phase ε_{Co} samples. A series of EDX analyses was performed from the edge to the interior of the foil and the results plotted in the form of the relationship:

$$I_{CoK\alpha}/I_{WL\alpha} = f(I_{CoK\alpha}.t)$$

where $I_{CoK\alpha}.t$ was proportional to the foil thickness t. The composition of the alloy was determined by atomic absorption analysis and the k factor was evaluated as

$$k_{W-Co} = 4.8 \pm 0.3$$

In order to explore the microchemical changes accompanying the growth process of the discontinuous precipitates high spatial resolution thin foil microprobe analyses were made by Zieba *et al.* (1997) across the moving cell boundary and across each set of several ε_{Co} and Co_3W lamellae.

As the ageing time was increased, the initial transformation product (identified as 'primary' lamellae with a spacing less than 10 nm) was gradually replaced by 'secondary' lamellae with a larger spacing, typically several tens of nanometres (Figure 5.20).

Figure 5.21 shows the results of an analysis trace made across the 'primary' lamellae/'secondary' lamellae interface shown in Figure 5.20 as the line X-X.

Figure 5.21 shows that the analysis of the 'fine' lamellae included both the W-rich and W-poor phases, and does not reveal any deviation from the original W-composition of the alloy. There is an abrupt change of solute concentration at the interface, consistent with the discontinuous mechanism of transformation.

Figure 5.22(A) shows a regularly spaced colony of 'secondary' cellular precipitates observed after 3 days ageing at 875 K, and Figure 5.22(B) shows the solute concentration profiles which were measured at distances of $X_1 = 85$ nm, $X_2 = 185$ nm and $X_3 = 310$ nm from the cell boundary as marked in Figure 5.22(A).

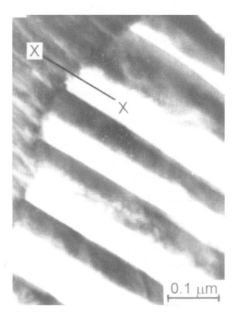

Figure 5.20. TEM micrograph of Co-32 wt%W alloy aged at 875 K for 3 days, showing fine 'primary' lamellae having been replaced by coarser 'secondary' lamellae. (Reproduced with permission by Zieba *et al.*, 1997).

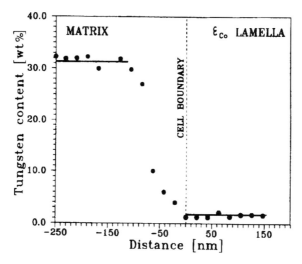

Figure 5.21.   Composition profile across the 'primary'/'secondary' lamellae boundary shown in
Figure 5.20 (line X-X). (Reproduced with permission of Zieba *et al.* 1997.)

The solute concentrations at the reaction front were not the equilibrium values
(the ε_{Co} and Co_3W solvus are marked in Figure 5.22(B)), in contrast to the value
approximately 300 nm from the reaction front.

Zieba *et al.* (1997) also made measurements of the W concentration within
individual ε_{Co} lamellae, by making analysis line scans as close as possible to the
reaction front (usually within 50–100 nm), Figure 5.23(A). The data shown in
Figure 5.23(B) are the average of three line scans.

5.3.6.6. *Mapping of Solute Segregation to Grain Boundaries.*

Before the advent of
EDS in the AEM, the only way to study grain boundary segregation was through
surface-chemistry techniques such as AES, SIMS and XPS, which require that the
segregated boundaries be brittle enough to fracture when loaded *in situ* in the
instrument. The creation of a partial or full monolayer of a different element
covering the grain boundaries can lead to such deleterious behaviour as temper
embrittlement in alloy steels, stress-corrosion cracking and other forms of
intergranular failure. Sometimes such segregation can be advantageous, as in the
case of sintering aids for ceramics and the de-embrittlement of intermetallics.

This kind of analysis is most effectively conducted in a dedicated FEGSTEM
instrument, which has an increased beam current while maintaining a small probe
size. Such instruments also permit the detection of light elements such as B at
boundaries. It has become possible to map the distribution of grain-boundary

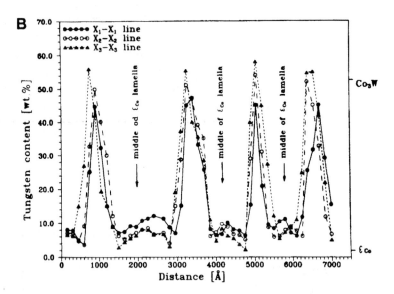

Figure 5.22. (A) TEM micrograph of Co-32 wt%W alloy aged at 875 K for 3 days showing a regularly spaced colony of 'secondary' precipitates. (B) Microanalytical scans across lamellae at three distances from the cell boundary. Equilibrium values of ε_{Co} and Co_3W are also marked. (Reproduced with permission by Zieba *et al.* 1997.)

segregants (Williams *et al.*, 1998), and to gain significant statistical information about the variation in the distribution of segregant from boundary to boundary.

Carpenter *et al.* (1999) prepared, by sputter depositing on NaCl, a 100 nm Al film alloyed with 4 wt% Cu. This they stabilized for 1 h at 475°C, then aged for 4 h at

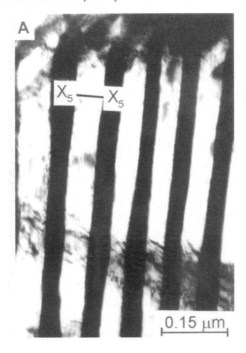

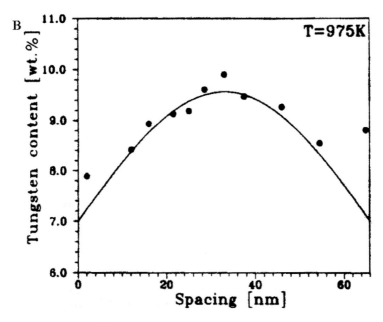

Figure 5.23. (A) TEM micrograph of Co-32 wt%W alloy aged 5 h at 975 K. (B) Line scan within ε_{Co} lamellae along X_5-X_5. (Reproduced with permission by Zieba *et al.* 1997.)

250°C to cause segregation of Cu to the grain boundaries. The specimen was removed from the substrate by dissolution and affixed to a Cu grid. The average grain diameter was 150 nm and the specimen was uniformly thin. The specimen was analysed in a FEGSTEM operating at 300 kV: at each pixel in the image, integrated $Al_{K\alpha}$ and $Cu_{K\alpha}$ peak intensities were collected, along with the Bremsstrahlung intensity from the background near each peak. Using image-processing software, the Bremsstrahlung intensity is subtracted from the peak intensity at each pixel in the image, after which the net X-ray intensities were converted into concentrations C_{Al} and C_{Cu} by use of a Cliff–Lorimer k factor (equation 1) experimentally determined from nearby particles of the θ-phase (assumed to have the stiochiometric composition of 33.3 wt% Cu).

Highest sensitivity to local concentration variations is achieved when the probe size is matched to the pixel size of the map. Figure 5.24(A) is a high magnification map of an edge-on grain boundary with 64×64 pixels, acquired at 2 MX with a probe size of ~ 1 nm FWTM and a probe current of ~ 0.5 nA. With a dwell time of 200 ms per pixel, the total frame time for this map was ~ 0.5 h.

Figure 5.24(B) shows a line profile extracted from the map of Figure 5.24(A) by averaging over 30 pixels parallel to the boundary direction corresponding to an actual distance of about 20 nm. The analytical resolution was ~ 4 nm, and the error bars (95% confidence) were calculated from the total Cu X-ray peak intensities (after background subtraction) associated with each data point in the profile (the error associated with Al counting statistics was assumed to be negligible). It is clear that these mapping parameters are not suitable for measurement of large numbers of boundaries, since typically only one boundary can be included in the field of view.

Maps gathered at lower magnifications (say < 500 kX) are useful for the systematic study of grain boundary segregation, since many boundaries can be analysed in a single field of view, while still maintaining high spatial resolution.

An example of a boundary analysed at a magnification of 230 kX by Carpenter *et al.* (1999) is given in Figure 5.25(A) and 5.25(B).

The mapping conditions were chosen by the authors to minimize the frame time and to maximize the counting statistics by increasing the probe size and probe current, while reducing the number of pixels in the map. These authors summarized the effect of the low-magnification mapping parameters on the measured analytical resolution for their specimen/microscope combination in Table 5.1.

5.3.6.7. Quantification of Grain Boundary Coverage. In addition to making comparisons of segregation levels between different boundaries, it is also possible to measure the fraction of boundaries showing different levels of segregation. Carpenter *et al.* (1999) used the same Al-4 wt% Cu specimen described above, and

Table 5.1. Low-magnification mapping parameters and measured analytical resolutions.

Magnification (kX)	Probe size (FWTM, nm)	Probe current (nA)	Frame time (hr)	Analytical resolution (FWTM, nm)
2000	~1	~0.5	0.5	4
240	~1.6	~0.9	5	4 (estimate)
300	~1.6	~0.9	1.5	4
230	~2.7	~1.9	0.75	8

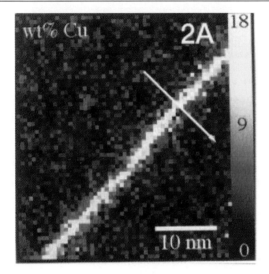

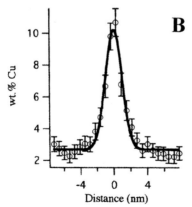

Figure 5.24. (A) Quantitative Cu map of an edge-on grain boundary in Al–4wt% Cu alloy foil. Magnification 2 MX. Composition range is shown on the intensity scale at RHS (Reproduced with permission by Carpenter *et al.* 1999). (B) Line profile extracted from Figure 5.24a averaged over 30 pixels parallel to the boundary. The solid curve is a Gaussian distribution fitted to the data (Reproduced with permission by Carpenter *et al.* 1999).

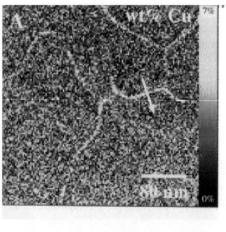

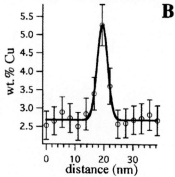

Figure 5.25. (A) Quantitative Cu map of an Al–4 wt% Cu film at 230 kX, 128 × 128 pixels, probe size
~2.7 nm, probe current ~1.9 nA, dwell time 120 msec per pixel, frame time ~0.75 hr. Composition range
is shown on the intensity scale (Reproduced with permission by Carpenter *et al.* 1999). (B) Line profile
extracted from the edge-on boundary marked in Figure 5.25a, averaged over 20 pixels (~55 nm) parallel to
the boundary, showing an analytical resolution of ~8 nm FWTM. Error bars represent 95% confidence,
and solid curve is a Gaussian distribution fitted to the data (Reproduced with permission by Carpenter
et al. 1999).

(assuming that all the copper is located at the grain boundary plane) the Cu
segregation (N_{Cu}) may be quantified in Cu atoms/nm^2 as:

$$N_{Cu} = \frac{A}{l_b} \frac{\rho_{Al} N_{Av}}{W_{Al}} (x_{Cu}^b - x_{Cu}^m)$$

where A is the area analysed, l_b is the length of grain boundary analysed, ρ_{Al} and
W_{Al} are the density and atomic weight of Al, N_{Av} is Avogadro's number, and x_{Cu}^b
and x_{Cu}^m are the atom fractions of Cu measured from the boundary and the matrix
respectively.

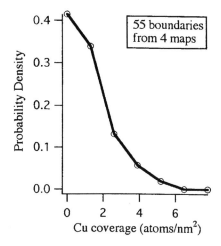

Figure 5.26. Distribution of Cu coverage (in atoms/nm^2) from 55 grain boundaries, extracted from four quantitative maps. (Reproduced with permission by Carpenter *et al.* 1999.)

N_{Cu} was calculated for 55 boundaries from four quantitative X-ray maps, and the probability distribution is shown in Figure 5.26. The mean coverage of this distribution is ~ 2.2 atoms/nm^2, and these data illustrate the feasibility of measuring segregation by X-ray mapping in the AEM in the case of fine-grained, thin film materials.

5.3.7 *Spatial Resolution and Detection Limits of Analytical STEM*

Williams *et al.* (2002) have reviewed the current state of AEM X-ray microanalysis, and they suggest ways in which the highest resolution of X-ray mapping may be achieved in the STEM with an EDS spectrometer. Because of their small collection angles and thin specimens, very small numbers of X-ray counts are generated, so the minimum detection limit is typically at best ~ 0.1 wt%. This value is an order of magnitude worse than the ~ 0.01 wt% figure for bulk-specimen in an SEM/EPMA.

The limited space available at the stage of a STEM means that only an EDS system can be interfaced with the microscope, providing an energy resolution of 120–140 eV, rather than the ~ 5–20 eV available with a WDS. The collection angle is at best 0.1 sr, and the best quantitative X-ray microanalysis is conducted using dedicated STEMs operating at 100 or 300 kV with cold-FEG electron sources. High spatial resolution X-ray maps may be constructed with spectral collection times as low as 50 ms per pixel, total mapping times as long as 80 min without drift correction, spatial resolution as low as 1.5 nm and detection limits of ~ 2 atoms in the analysed volume.

Ziebold (1967) has defined the detection limit of any analytical technique in terms of the minimum mass fraction (MMF) detectable according to the expression:

$$\text{MMF} = \frac{a}{\sqrt{P(P/B)\tau}}$$

where a is a constant (~ 1), P is the X-ray peak intensity, B is the background intensity, and τ is the analysis time. Williams *et al.* (2002) observe that all three factors of P, P/B and τ should be increased in order to improve the analytical sensitivity in thin-specimen microanalysis by AEM. Brighter guns than the cold FEG are not expected in the near future, and thicker specimens degrade the minimum detectable mass (MDM) (i.e. the number of detectable atoms in the analysed volume, even though P is increased. The only option left to increase P is to improve the X-ray detection efficiency, and these authors draw attention to the existence of a totally new spectrometer (called the bolometer XEDS detector) (Wollman *et al.*, 2000), based on microcalorimeter technology that detects the temperature change by a superconducting transition sensor when a single X-ray photon is absorbed in a Ag foil cooled to $\sim 100\,\text{mK}$. The energy resolution of the bolometer XEDS is $< 10\,\text{eV}$, which is comparable to that of WDS detectors.

Williams *et al.* (2002) have plotted the relationship between the spatial resolution and the *MMF* in several electron-probe instruments, and this is reproduced in Figure 5.27.

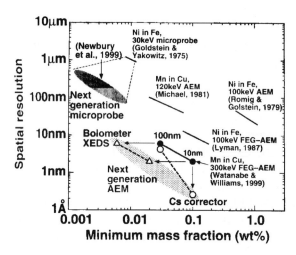

Figure 5.27. A summary of the relationship between the spatial resolution and the MMF for X-ray microanalysis in several electron-probe instruments (references given). The two shaded areas represent the ranges of future EPMA microanalysis predicted by Newbury *et al.* (1999) and the future AEM microanalysis estimated by Williams *et al.* (2002). (Reproduced by permission of Williams *et al.*, 2002.)

In addition to the data from previous and current instruments, predictions of the next generation X-ray microanalysis are also plotted as the shadowed areas of Figure 5.27. Newbury *et al.* (1999) have estimated that X-ray microanalysis of bulk samples becomes possible at 5 keV or lower accelerating voltages if a FEG-EPMA is developed capable of interfacing to the bolometer XEDS system. The lower accelerating voltages will make the analysed volume smaller, and hence the spatial resolution can approach the several hundred-nm levels. The bolometer detectors have the potential to resolve the low energy X-ray (M or N) lines, so trace-element analysis will reach ~ 0.001 wt% (10 ppm) levels.

With regard to X-ray microanalysis in the AEM, the bolometer detectors can lower the detectability limits from 0.03 wt% to ~ 0.006 wt% in 100 nm thick Cu-Mn specimens. Even in 10-nm-thick specimens, ~ 0.02 wt% will be detectable with these detectors.

5.4. AUGER ELECTRON SPECTROSCOPY (AES)

AES was developed in the late 1960s, and in this technique electrons are detected after emission from the sample as the result of a non-radiative decay of an excited atom in the surface region of the sample. The effect was first observed in bubble chamber studies by Pierre Auger (1925), a French physicist, who described the process involved.

It is a surface specific technique utilising the emission of low energy electrons, and is one of the most commonly employed analytical techniques for determining the composition of the surface layers of a material sample. AES has a number of advantages over EPMA: the technique offers high sensitivity of analysis, typically of the order of 1% of a monolayer for all elements except H and He, being most sensitive to the elements of low atomic number, and it may be employed as a means of monitoring the surface cleanliness of samples. The basic technique has been adapted for use in Auger Depth Profiling, which provides quantitative compositional information as a function of depth below the surface. Scanning Auger Microscopy (SAM) has also been developed, which provides spatially resolved compositional information on heterogeneous samples.

The methods of sample preparation for AES are identical to those used in XPS (see Chapter 2.1), the objective again being to ensure that the surface to be analysed has not been contaminated or altered prior to analysis. AES must be carried out in UHV conditions.

An electron gun (see Section 5.1.3) provides the requisite electron source for AES, and may consist of a tungsten or a LaB_6 cathode, or a Field Emission source. The latter provide the brightest beams, and beam widths as narrow as 10 nm permit

Auger analysis of small features. The primary electron beam column is similar to that in electron microscopes, and it may contain both electrostatic and magnetic lenses for beam focussing as well as quadrupole deflectors for beam steering and octopole lenses for beam shaping.

The physical basis of AES involves three basic steps, namely *atomic ionization* (by the removal of a core electron), *electron emission* (the Auger process), and *analysis of the emitted Auger electrons*.

5.4.1   *The Emission of Auger Electrons from Ionized Atoms*

The Auger process is somewhat more complicated than that of X-ray photoemission (see Section 5.1.2). Let us firstly consider the energies of the various energy levels in an isolated, multi-electron atom (Figure 5.28).

The conventional chemical nomenclature for these orbitals is given on the right hand side, and the X-ray nomenclature used in Auger spectroscopy appears on the left. The designation of levels to the $K, L, M, \ldots$ shells is based on their having principal quantum numbers of $1, 2, 3, \ldots$ respectively.

Levels with a non-zero value of the orbital angular momentum quantum number $(l > 0)$, i.e. $p, d, f, \ldots$ levels, show spin-orbit splitting. The magnitude of this splitting is too small to be evident in Figure 5.28, hence the double subscript for these levels (i.e. $L_{2,3}$ represents both the L_2 and L_3 levels.

In the solid state the core levels of atoms essentially remain as discrete, localized levels as shown in Figure 5.28. The valence orbitals overlap significantly with those of neighbouring atoms, generating *bands* of spatially delocalized energy levels.

The Auger process is initiated by the creation of a core hole, which is typically carried out by exposing the sample to a beam of high energy electrons (with a typical

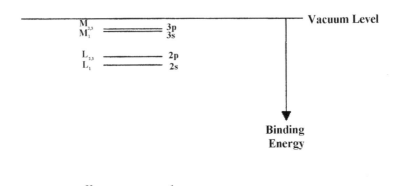

Figure 5.28.   Schematic illustration of the various energy levels in an isolated atom.

primary energy in the range 2–10 keV). Such electrons have sufficient energy to ionize all levels of the lighter elements, and higher core levels of the heavier elements.

The ionized atom that remains after the removal of the core hole electron is in a highly excited state and will rapidly relax back to a lower energy state by one of two routes, namely *X-ray fluorescence* (Section 5.1.2) or by transferring the energy to an electron in another orbit, which, if it has sufficient energy, will be ejected into the vacuum as *Auger emission*. An example of the latter process is illustrated in Figure 5.29.

In the example illustrated in the diagram, the atom is raised into an excited state by the creation of a core hole at level L_3. An electron then falls down from a higher level (level M_1 in the diagram) to fill this core hole, and the excess energy is carried away as the kinetic energy of a further electron which is emitted from the atom (in this case from level $M_{2,3}$).

X-ray nomenclature is used for the energy levels involved, and the Auger electron is described as originating from an 'XYZ' Auger transition, where X is the level of the original core hole, Y is the level from which the core hole was filled, and Z is the level from which the Auger electron was emitted. This last electron, the so-called XYZ Auger electron, thus has a well-defined energy. In the example of Figure 5.29, the Auger transition illustrated is described as $L_3M_1M_{2,3}$. The final state is a doubly ionized atom with core holes in the M_1 and $M_{2,3}$ shells.

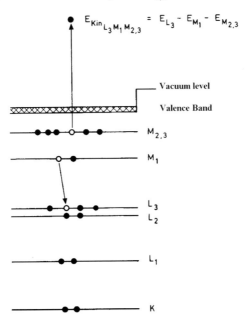

$$E_{Kin_{L_3 M_1 M_{2,3}}} = E_{L_3} - E_{M_1} - E_{M_{2,3}}$$

Figure 5.29. An energy level diagram showing the physical basis of the Auger technique.

The kinetic energy of the Auger electron emitted in the example of Figure 5.29 (E_{Kin}) can be estimated from the binding energies of the various levels involved:

$$E_{Kin} = E_{L_3} - E_{M_1} - E_{M_{2,3}} - \phi \tag{1}$$

where ϕ is the work function (not included in the diagram of Figure 5.29).

The second and third energy terms in equation (1) could be interchanged without any effect (i.e. it is impossible to say which electron fills the initial core hole and which is ejected as an Auger electron: they are indistinguishable. The existence of different electronic states within the final doubly ionized atom may furthermore lead to fine structure in high-resolution spectra.

Since the initial hole may be in various shells of a given atom, there will be many possible Auger transitions for a given element, and of varying intensity. Auger spectroscopy is based upon the measurement of the kinetic energies of the emitted electrons – each element in a sample being studied will give rise to a characteristic spectrum of peaks at various kinetic energies.

Originally, attempts were made to estimate the modifications to E_{M_1} and $E_{M_{2,3}}$ arising from the excited nature of the emitting atom, in order to assist in the calculation of expected Auger energies. Today, Auger spectra are generally used as a fingerprint for identifying elements present in the surface region of the sample by comparing the spectrum from the unknown with standard spectra of the elements from one of the reference handbooks.

5.4.2 Instrumentation

The energies of the Auger electrons leaving the sample are determined in a manner similar to that employed for photoelectrons already described in chapter 2 Section 4. Modern instruments nearly always incorporate cylindrical mirror analysers (CMA) because their high transmission efficiency leads to better signal-to-noise ratios than the CHA already described.

Figure 5.30 illustrates schematically the cross-section of a CMA, and the principle of its operation. It consists of two coaxial cylinders, with the inner cylinder at ground potential and a potential of $-V$ on the outer. The primary electron beam hits the sample surface and some of the Auger electrons generated will pass through the grid covered annular aperture in the inner cylinder.

A variable negative potential on the outer cylinder bends the Auger electrons of a particular kinetic energy E_0 back through a second annular aperture on the inner cylinder; they are then are focussed at an exit aperture on the analyser axis where they are collected by an electron multiplier. The energy of the transmitted electrons is proportional to the voltage on the outer cylinder (V), and simply scanning the

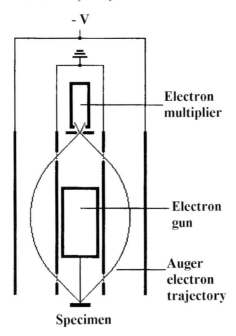

- V

Electron
multiplier

Electron
gun

Auger
electron
trajectory

Specimen

Figure 5.30. Schematic diagram of a cylindrical mirror electron energy analyser.

potential $-V$ on the outer cylinder of a CMA gives directly the energy distribution of electrons passing through it.

5.4.3 Chemical Analysis

When the electron beam enters the sample, it penetrates a small volume, typically about one cubic micron ($10^{-18}\,m^3$). X-rays are emitted from most of this volume, but Auger signals arise from much smaller volumes, down to about $3 \times 10^{-25}\,m^3$. The Auger analytical volume depends on the beam diameter and on the escape depth of the Auger electrons. The mean free paths of the electrons depend on their energies and on the sample material, with values up to $\sim 25\,nm$ under practical analytical conditions.

5.4.3.1. Qualitative Analysis. The Auger electrons start with narrow energy distributions, but they lose energy as they pass through material. They will fail to emerge with their characteristic energies if they start from deeper than 1 to 5 nm beneath the surface: such Auger electrons that escape from deeper in the sample contribute loss tails to the spectrum background. The sum of these and other interfering signals (arising from secondary electrons generated by inelastic scattering processes) is much greater than the Auger signals themselves, so the Auger spectrum data are often displayed in *differentiated* form in order to enhance the signal relative to the interference. This approach is illustrated in Figure 5.31, due to Flewitt and Wild (1985), for an Auger spectrum from a stainless steel surface.

This spectrum was obtained from the surface of a sample of an iron-chromium-nickel alloy, and peaks from these elements, resulting from LMM transitions, are identified together with those from contaminants such as carbon, sulphur and oxygen.

Today, with the increasing use of computers to record and process spectra, the problem of background removal and the display of the Auger electron peaks in the direct mode is relatively straightforward. For this reason the use of the direct mode has recently become more popular.

Since the electron binding energies change when the atom forms a compound, the Auger peak energy also changes, so that the chemical state of the elements at the surface can be identified by these *chemical shifts*, provided a high-resolution energy analyser is employed.

Auger 'atlases' can be used in applied AES for rapid preliminary elemental identification, although comparison of the spectra from any two adjacent elements in such atlases reveals that the relative intensities of different Auger transitions within

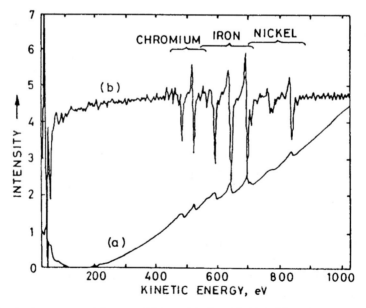

Figure 5.31. An Auger spectrum from a stainless steel surface: (a) the undifferentiated N(E) mode, (b) the differentiated mode. (After Flewitt and Wild 1985.)

the same element show large variations. There is no simple semi-empirical formula that will produce a set of usable relative intensities.

5.4.3.2. *Quantitative Analysis.* In its basic form, AES provides compositional information on a relatively large area ($\sim 1 \, \text{mm}^2$) of surface, using a broad-focussed electron beam probe. Sufficient signal may be obtained in this way with a low incident electron flux, thus avoiding potential electron-induced modifications of the surface.

A detailed review of the quantification of AES is given by Seah in Briggs and Seah (1990), who points out that an analyst would in practice blend the three approaches of (a) calculation of all the relevant terms from first principles, (b) the use of locally produced standards and databases, and (c) the use of published databases.

Considering first the case of *homogeneous binary solids* consisting of a matrix M in which atoms A are embedded:

The ionization cross-section, $\sigma_x(E)$, of level X by electrons of energy E, contains contributions from the primary electron beam (of energy E_0) with additional ionization arising from back-scattered electrons. The back-scattering term, r_M, is dependent on the matrix M in which the A atoms are embedded and the angle α to the surface normal of the incident electron beam. A final term giving additional ionization of the core level X arises from transitions between shells of the same principal quantum number (Coster-Kronig transitions): the more weakly bound levels have added ionization and give stronger Auger electron peaks.

The total contribution to the Auger electron signal is then dependent upon the *attenuation length* (λ_M) in the matrix before being inelastically scattered, and upon the transmission efficiency of the electron spectrometer as well as the efficiency of the electron detector. Calculated intensities of Auger peaks rarely give an accuracy better than $\sim 50\%$, and it is more reliable to adopt an approach which utilises standards, preferably obtained in the same instrument.

Quantification of inhomogeneous samples is a more usual situation encountered in the application of AES. Figure 5.32 (after Seah 1986) show some of the common configurations of elements that may be studied by AES (or XPS).

Modern spectrometers only require electron beam currents in the range 0.1–10 nA and hence probe sizes of 20–200 nm may be readily achieved with thermionic sources and 5–15 nm with a FEG. Spatially resolved compositional information on heterogeneous samples may be obtained by means of the *Scanning Auger Microprobe (SAM)*, which provides compositional maps of a surface by forming an image from the Auger electrons emitted by a particular element.

SAM is a combination of the techniques of SEM and AES: an electron beam of high energy (3–10 keV) is scanned over the surface and the electrons excited from the

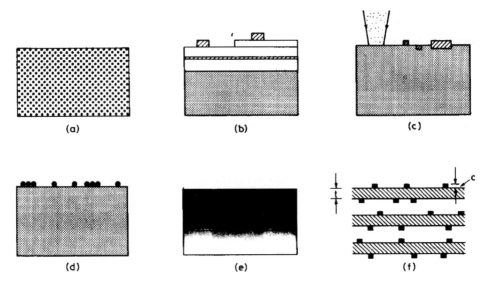

Figure 5.32. Illustrating common configurations of elements that may be quantified by AES analysis :
(a) homogeneous (b) large, well-defined structures varying laterally and with depth (c) small, well-defined
structures varying laterally and with depth, (d) monolayer segregants and absorbed layers (e)
layered composition gradients and (f) a catalyst promoter on a layered support. (After Seah 1986,
Crown Copyright.)

surface are energy analysed to detect Auger peaks. Again, all elements in the periodic
table (except hydrogen and helium) can be detected, and depth analysis is in the
range 3–5 nm compared to the 1–3 microns analysis depth of EDX.

The Physical Electronics 680 Nanoprobe employs a field emission electron gun,
and this results in a spatial resolution of less than 10 nm. Ion bombardment for
depth profiling is available in the SAM, and both the electron beam and the ion
beam are computer controlled so that depth profiles can be run automatically, and
maps and line scan of Auger electron distributions can be generated.

Complete characterisation of complex samples thus becomes possible, and the
SAM is used to identify embedded defects and surface particles in semiconductor
devices, as well as the study of metal matrix composites and grain boundary analysis.

The total Auger emission rises rapidly as the angle of incidence of the electron
beam α is increased, and this leads to a change in intensity with topography in
imaging. To make the image intensity approximately quantitative, Prutton *et al.*
(1983) have proposed the function:

$$T = \frac{N_1 - N_2}{N_1 + N_2}$$

where N_1 is the signal on an Auger electron peak and N_2 is that on the background at slightly higher energy, and its use largely removes topographical effects.

Briggs and Seah (1990) consider in detail a number of the situations illustrated in Figure 5.32. For example:

5.4.3.3. Monotonic Overlayers.

In the absence of elastic scattering, the signal of the substrate B covered by a fractional monolayer ϕ_A of A is

$$I_B = I_B^\infty \{1 - \varphi_A + \varphi_A \exp[-a_A/\lambda_A(E_B) \cos \theta]\} \tag{2}$$

The signal from the overlayer is

$$I_A = \phi_A I_A^\infty \left[\frac{1 + r_B(E_A)}{1 + r_A(E_A)} \right] \{1 - \exp[a_A/\lambda_A(E_A) \cos \theta]\} \tag{3}$$

Because of sample roughness and all other changeable settings, *intensity ratios* such as I_A^∞/I_B^∞ are employed, so that fractional monolayer coverage ϕ_A is given by:

$$\frac{\varphi_A \{1 - \exp[-a_A \lambda_A(E_A) \cos \theta]\}}{1 - \varphi_A \{1 - \exp[-a_A/\lambda_A(E_B) \cos \theta]\}} = \left[\frac{1 + r_A(E_A)}{1 + r_B(E_A)} \right] \frac{I_A/I_A^\infty}{I_B/I_B^\infty} \tag{4}$$

The attenuation lengths and the back-scattering terms may be calculated from appropriate equations.

The quantification of *thicker overlayers*, and of *sputter-depth profiles* are also considered by Seah in the book referred to above. The latter technique involves a controlled surface etching of the surface region being analysed, while continuing to monitor and record the Auger spectra. This controlled surface etching is accomplished by simultaneously exposing the surface to an (argon) ion flux, at between 500 eV and 5 keV. This leads to sputtering (i.e. removal) of the surface atoms.

5.4.4 Areas of AES Application to the Surface Analysis of Materials

5.4.4.1. Insulating Materials.

Sample charging can occur during the Auger analysis on insulating materials. A small surface charge can cause the spectral lines to shift in energy or shift to several energies, thus broadening the peak, making peak height measurements unreliable. A large surface charge can make it impossible to obtain any spectra at all.

Sample charging can be avoided by mounting the sample at a grazing angle with respect to the electron beam. This increases the secondary electron emission from the sample, so that, in combination with a lowered beam energy and lowered beam current, the problem can be overcome. Some instruments are able to generate high energy electron beams (20 or 25 kV), and if the insulating layer being analysed is thin (1 to 1.5 μm) over a conducting layer (a common situation is some microelectronic devices), such a beam at normal incidence will create an ionized conduction path through the insulator to the conductor, and the analysis is straightforward.

Another approach is to hold the insulating sample in place by means of a thin metal plate or foil which has a small (3 mm) hole in it. This will provide improved electrical contact near the analysis area and minimizes the exposed area of insulating material.

5.4.4.2. Particle Identification.

In AES and SAM it is possible to steer the beam to a particular point and focus it to achieve very high spatial resolution (< 20 nm). The analysis volume is determined by the spot size of the incident beam and the inelastic mean free path of the Auger electrons. The latter varies from about 1 nm to about 5 nm for Auger electrons with energies from 30 eV to about 2 keV, and its magnitude is independent of the electron beam energy or beam current. Thus in the case of a surface particle of diameter, say, 0.5 μm, all of the Auger electrons would emanate from the particle and none from the adjacent area or substrate.

5.4.4.3. Characterisation of Segregation.

For small levels of the segregation ϕ_A of element A expressed in monolayers, equation (4) may be written:

$$\phi_A = Q_{AB} \frac{I_A/I_A^\infty}{I_B/I_B^\infty} \tag{5}$$

where I_A and I_B are the Auger intensities of the segregant and matrix, I_A^∞ and I_B^∞ are the intensities of the pure bulk materials and Q_{AB} is a matrix factor. Equation (5) may be further simplified to

$$\phi_A = m K_{AB} \frac{I_A}{I_B} \tag{6}$$

where $m = 1$ for surfaces and 2 for grain boundaries, and

$$K_{AB} = \left[\frac{\lambda_A(E_A) \cos \theta}{a_A} \right] \left[\frac{1 + r_A(E_A)}{1 + r_B(E_A)} \right] \frac{I_B}{I_A}$$

Here it is assumed that upon fracture of a grain boundary, half of the segregant is left on each free surface, and that the segregant atoms pack at the same density as they would in the pure elemental state.

5.4.4.4. Measurement of Grain Boundary Segregation.

An example of the application of AES to that study of segregation of a solute element to grain boundaries is provided by the work of Heatherly and George (2001) for the case of iridium containing 0.006 weight fraction thorium which segregates to the grain boundaries where it enhances cohesion. A series of samples were prepared which contained controlled amounts of a number of alloying elements, which were added at levels ranging from 50 to 5000 ppm.

Auger specimens $\sim 10 \times 2 \times 1$ mm were heat-treated and then notched to control the location of fracture. Two of these notched samples at a time were loaded into specimen grips such that they could be broken independently under UHV and then examined by a SAM instrument. Scanning electron micrographs allowed different grain boundaries and transgranular cleavage areas to be selected for analysis.

The size of the analysed areas was of the order $20 \times 20 \, \mu m$, and approximately 10 grain boundaries were analysed on each fractured sample, using a 5 KeV electron beam and a current of $\sim 150 \, nA$. The peak heights were measured on the differentiated spectra and their atomic concentrations were calculated using the appropriate Auger sensitivity factors for each element. The latter were calculated such they would give the correct bulk concentrations on analysing the transgranular spectra.

For the alloys doped with the impurities Fe, Ni, Cr and Al, the amounts of these elements (and Th) are plotted in Figure 5.33(a–d), and the undoped alloy is shown in Figure 5.33(e).

Each data point represents Auger results from a single grain boundary or transgranular region. It is seen that the impurities Fe, Ni, Cr and Al do not segregate to the grain boundaries (their intergranular and transgranular concentrations are similar), whereas thorium segregates strongly to the grain boundaries in all four alloys to a degree which is similar in the undoped alloy (Figure 5.33(e)). The lack of segregation is consistent with the relatively small size misfits with Ir and large solubilities in Ir of Fe, Ni, Cr and Al. The strong segregation of Th is consistent with its large size misfit and low solubility in Ir.

It is apparent that the degree of segregation of Th to the Ir grain boundaries varies from grain to grain, and this is to be expected, since Suzuki *et al.* (1981) have demonstrated that the amount of solute segregation at a grain boundary is strongly dependent upon the crystallographic orientation of the grain boundary plane.

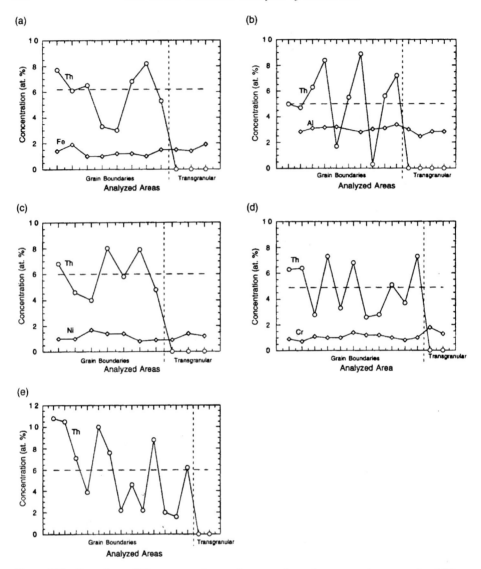

Figure 5.33. Impurity and Th concentrations on intergranular and transgranular areas of an iridium alloy containing 6000 ppm Th and (a) 5000 ppm Fe, (b) 4000 ppm Al, (c) 3000 ppm Ni, (d) 5000 ppm Cr and (e) no added impurities. (After Heatherly and George 2001).

Suzuki *et al.* studied phosphorus segregation in iron by AES using a similar technique to that described above. Additionally, the orientation of each grain relative to the specimen axis was determined by means of selected area channelling patterns (SACP), and the crystallographic orientation of the boundary plane was

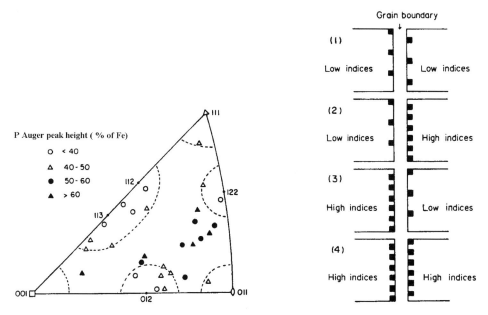

Figure 5.34. (a) Shows the relation between Auger peak height ratio (P/Fe) and the crystallographic orientation of grain boundary planes. It is apparent that boundaries of high misorientation exhibit the greatest P segregation. (b) Schematic illustration showing combinations of grain boundary planes with low index and high index: dark marks on grain boundary planes indicate segregated phosphorus. (After Suzuki *et al.* 1981.)

thus determined with respect to each of the component grains of the boundaries studied. Some of their data appear in Figure 5.34.

Figure 5.34(a) shows the relation between the amount of segregated P and the crystallographic orientation {hkl} of the grain boundary planes.

Here, the boundary plane index, {hkl}, is with respect to the grain analysed by SAM. The phosphorus segregation is found to be larger for the boundary planes with high indices. On the planes close to the low index planes, e.g. {012}, {112}, the degree of segregation is small. The situation is summarized in Figure 5.34(b), and the authors conclude that the true degree of segregation should be measured by AES at the boundary surface of both component grains in order to gain a true insight into the situation.

5.4.4.5. Intergranular Corrosion. In order to identify the species present in the narrow crevices of intergranular corrosion paths in metallic materials, which may be only 1–2 µm wide, the use of an Auger map using SAM has been successful.

Figure 5.35 is an image of an archaeological specimen comprising a small fragment of a Urartian bronze belt of about the 8th–7th centuries BC.

One surface of the fragment of bronze was ground and polished flat, so that a section through the internal corrosion below the surface was exposed. The polished surface was sputtered with an ion gun to clean it by the removal af about 3 nm of material. Figure 5.35(a,b) illustrate the distributions of copper and tin in a 20 μm square, the corrosion tracks being dark in Figure 5.35(a) (depleted in copper) and bright in Figure 5.35(b) (enriched in tin).

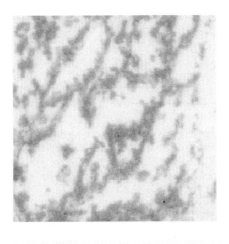

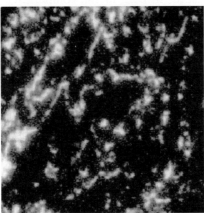

Figure 5.35. (a) SAM map for Cu, 20 μm square, of internal corrosion i bronze belt. (b) SAM map for Sn, 20 μm square of part of the same area as Figure 5.35(a). (Courtesy Dr J. P. Northover, Oxford University.)

Figure 5.36 shows a line scan obtained for these elements, as well as for oxygen, zinc and sulphur.

5.4.4.6. *Depth Profiling.* The quantitative interpretation of AES data requires a knowledge of the distribution of the grain boundary segregants both laterally on the

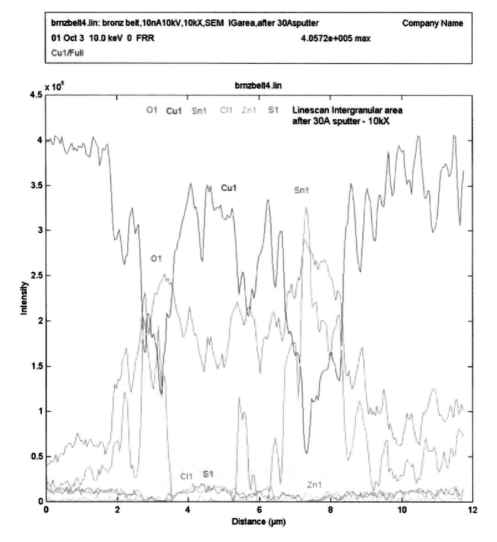

Figure 5.36. SAM line scan across an intergranular corrosion trace in bronze belt of Figure 5.35. (Courtesy Dr Scott Lea, Pacific Northwest Laboratories.)

fracture surface and with depth into the surface. The distribution with depth may be determined by AES with ion beam milling, and Figure 5.37, due to Seah (1980) illustrates the grain boundary segregation of P, Cr and Mo at the grain boundaries of a temper brittle 3Cr-0.5Mo rotor steel. Here 600 eV argon ions were employed to obtain the sputter profiles.

The profiles for most segregants, characterised by a rapid exponential decay with depth etched, are compatible with a single atom layer of segregant atoms at the fracture plane. The decay of the Auger electron intensity, I^A, for the sputtering of atoms at the fracture plane is described by :

$$I^A = I_0^A \exp(-SJAt),$$

where I_0^A is the initial intensity, S is the sputter yield of the segregant, J is the flux of ions per unit area, A the area of the sputtered ion, and t is the time elapsed. The exponential decay exhibited by the curve for phosphorus in Figure 5.37 thus

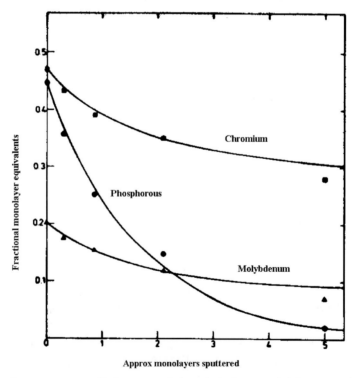

Figure 5.37. Average sputter profile from ten grain boundaries in temper brittle rotor steel using 600 eV argon ions. (After Seah 1980, Crown Copyright.)

indicates a monolayer segregation at the boundary for this element. For the other alloy constituents of the steel, Mo and Cr, the residual signals after long sputtering times are attributable to grain boundary carbides, but the initial falls in the signals again indicate grain boundary segregation of these elements.

In the field of microelectronics, there is continuing research in developing new materials to be used in semiconductor fabrication. They must be formed as thin films in a controlled, reproducible and uniform manner to be useful in semiconductor manufacturing applications. Depth profiling by AES is used to assess the properties of such films. The samples are sputtered with an argon ion beam and analysis performed using standard sensitivity factors, and it is possible to demonstrate that such films are uniform throughout a depth of, say, 250 nm.

5.5. ELECTRON ENERGY LOSS SPECTROSCOPY (EELS)

There are two types of electron energy loss spectroscopy currently in use. The first of these is found in scanning transmission electron microscopes. As indicated in Figure 5.1, compositional information may be obtained in the TEM by measuring the energy loss of the inelastically scattered electrons transmitted through a thin specimen.

The other technique is HREELS (high resolution EELS) which utilises the inelastic scattering of low energy electrons in order to measure vibrational spectra of surface species. The use of low energy electrons ensures that it is a surface specific technique, and is often chosen for the study of most adsorbates on single crystal substrates.

We will deal in the greater detail with the former application.

5.5.1 The Physical Processes

Inelastic scattering occurs as a result of Coulomb interaction between a fast incident electron and the atomic electrons surrounding each nucleus. *Single electron excitation* and *plasmon excitation* represent alternative modes of inelastic scattering. In the former process a fast electron may interact with an *inner shell* electron, leaving the target atom in an ionized state, and it will quickly lose its excess energy as X-rays or Auger emission as an electron of lower binding energy undergoes a downward transition to the vacant 'core-hole'. An *outer shell* electron can also undergo single-electron excitation, and may be emitted as a secondary electron. In plasmon excitation, outer shell inelastic scattering may involve many atoms of the solid (known as a plasma resonance), so the excess energy is shared among many atoms, and in the decay (deexcitation) process energy is released in the form of heat.

The secondary processes of electron and photon emission can be studied in detail by appropriate spectroscopies, and we have discussed these elsewhere. In EELS we study directly the primary processes of electron excitation, each of which results in a fast electron losing a characteristic amount of energy. The beam of transmitted electrons is directed into a high-resolution electron spectrometer which separates the electrons according to their kinetic energies – producing an electron energy loss spectrum, showing the scattered intensity as a function of the decrease in kinetic energy of the fast electron.

The interaction of a high-energy electron with an atom involves an exchange of momentum and of energy. Since momentum is a vector quantity a complete description of the process requires a knowledge of the angular displacement θ as well as the energy change ΔE, as illustrated in Figure 5.38 (Joy 1981).

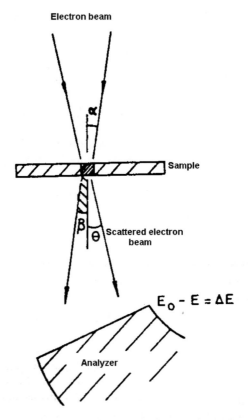

Figure 5.38.   Illustrating the incident electron beam convergence angle (α), the scattering angle (θ) and the spectrometer acceptance angle (β) in electron energy loss spectroscopy. (After Joy, 1981.)

The signal $I(E)$ detected at some energy loss E by a spectrometer collecting electrons scattered through angles up to β is given by:

$$I(E) = I.N.\eta.\sigma(\beta, E, E_0) \tag{1}$$

where I is the intensity of the incident electron beam of energy E_0, N is the number of atoms in the irradiated area, η is an efficiency factor and σ is the interaction cross section (which represents the probability that any incident electron suffers an energy loss E while being scattered into a solid angle less than β).

5.5.2 Instrumentation for Energy-Loss Spectroscopy

Egerton (1986) gives a detailed account of the types of spectrometers that may be used for electron energy loss analysis made in conjunction with transmission electron microscopy. In the microscope column, the spectrometer must be mounted after the specimen and is usually the final component of an analytical microscope. For good electron transmission an incident energy of $\sim 10^5 \, \text{eV}$ is required, and the desired energy resolution is $\sim 1 \, \text{eV}$, so an instrument of high resolving power is essential.

A magnetic prism spectrometer is usually employed, in which a magnetic field is used to deflect all the electrons through about 90 degrees, as illustrated in Figure 5.39.

The more energetic electrons will be deflected through a slightly smaller angle than those of less energy, and so they will be dispersed into a spectrum of energies. There are two ways of detecting the spectrum. In a *serial spectrometer*, the spectrum is scanned across a slit, by varying the strength of the magnetic field, so that each energy is detected in turn by the single detector. In a *parallel spectrometer* a multiple 'position-sensitive' detector is employed, and the whole spectrum can be collected at once, the technique being known as PEELS (parallel electron energy-loss spectrometry). The PEELS method is in dominant use, since it is intrinsically more efficient, as it counts all energies all of the time. Three parameters define the performance of this design of spectrometer, the dispersion (which is related to the radius of curvature of the spectrometer and is typically a few $\mu m/eV$ for $E_0 = 100 \, \text{keV}$), the solid angle acceptance ($\pi\beta^2$) and the energy resolution which is typically a few eV.

5.5.3 Energy Loss Spectra

A typical loss spectrum is recorded over a range of about 1000 eV, and is conventionally considered to consist of three regions. The first *zero-loss*, or 'elastic' peak represents electrons which are transmitted without suffering any measurable energy loss. The *low loss region*, containing electrons which have lost up to about

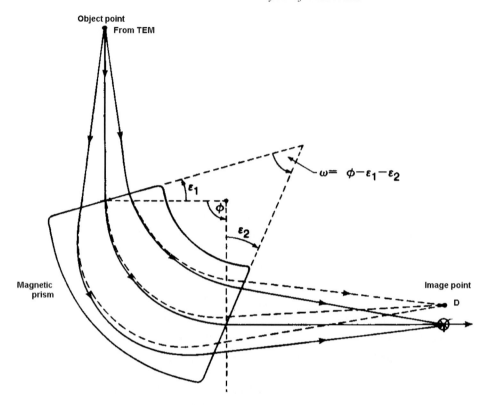

Figure 5.39. Illustrating the focussing and dispersive properties of a magnetic prism in an electron energy-loss spectrometer. In a *serial* spectrometer a slit at D is used to ensure that only electrons of a single energy loss enter the detector. In a *parallel* spectrometer, a position-sensitive detector is placed at D to collect electrons of all energies in parallel.

50 eV, arises largely from plasmon scattering. The plasmon peaks are not very useful for analysis because the energies at which they occur are similar for many materials.

In the third region, at higher energy loss, the electron intensity decreases according to some fairly high power of energy loss, making it convenient to use a logarithmic scale for the recorded intensity. A minimum energy E_k is needed to ionize a particular inner shell of an atom, and for $E > E_k$ a finite cross-section of ionization exists to increase $I(E)$ when this condition is satisfied. This will produce a discontinuity or 'edge' in the spectrum at E_k. Since E_k approximates to the binding energy of the particular atom shell ionized, it uniquely characterises the atom and a single measurement of the energy loss identifies the element. It is these edges which are routinely used for EELS analysis.

The drawing of Figure 5.40 illustrates the three regions referred to above, namely a zero-loss peak, a low-loss peak and a series of characteristic edges corresponding

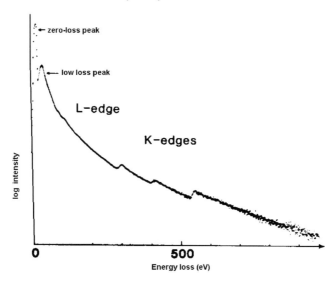

Figure 5.40. Illustration of an electron energy-loss spectrum showing the three typical regions: a zero-loss peak, a low-loss peak and L and K edges.

to inner shell excitation. Qualitative analysis is carried out by determining the energies of any edges which can be seen, and comparing them with tabulated values for the elements. One of the great strengths of the technique is that it can be used to detect elements whose X-rays are too soft to be counted: it is possible to detect EELS edges from He, Li and Be, for example, which are otherwise very difficult to analyse.

5.5.3.1. Quantitative Analysis. Quantitative elemental analysis is possible by measuring an area under the appropriate ionization edge, making allowance for the background intensity.

The background is substantial under the edges to be measured: the cross-section term, σ, in equation (1) contains contributions from other interactions, together with the required edges that combine to produce a signal:

$$I(E) = I.A.E^{-r} \tag{2}$$

where A and r are constants for limited ranges of the spectrum whose values depend upon E, β and the material. Peak deconvolution procedures are necessary to obtain a quantitative measure of elemental concentration. Equation (2) is used to fit the background shape at a position on the spectrum before the edge position.

Thus from equation (1), if the intensity at an edge is I_k, then

$$N = I_k/I\sigma\eta \tag{3}$$

In practice, the area under the peak can be obtained for an energy window, Δ, as indicated in Figure 5.41, so that:

$$N = I_k(\beta, \Delta)/I\sigma(\beta, \Delta)\eta \tag{4}$$

In order to allow for back scattering and plasmon scattering in the sample to be evaluated directly from the spectrum, I in equation (4) is conventionally replaced by $I_0(\beta\Delta)$, the integral in the energy window under the zero-loss peak, for the acceptance angle β. This permits the calculation of the absolute number of atoms which contributed to the edge.

Elemental ratios are usually required, i.e.:

$$N_A/N_B = I_{kA}(\sigma_A\eta_A)/I_{kB}(\sigma_B\eta_B) \tag{5}$$

Giving mass concentration ratios:

$$C_A/C_B = (A_A N_A/t\rho)/(A_B N_B/t\rho) = (A_A N_A)/(A_B N_B) \tag{6}$$

where t is the sample thickness, ρ is specimen density and A_A the atomic weight of element A.

Using this approach the relative atomic abundances of light elements in the specimen are easily determined.

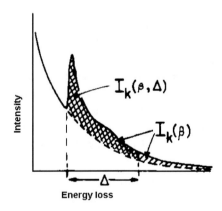

Figure 5.41. Schematic diagram of the low loss region and an ionization edge recorded with an angle-limiting collection aperture of semi-angle β.

5.5.3.2. The Fine Structure Before and After Each Edge. ELNES is the term use to describe the energy-loss near edge structure, and this can be quite different for an element in different compounds. For example the shape of the aluminium L edges are quite different in EELS spectra from metallic aluminium and aluminium oxide, so that the chemical form of a given element may be indentified from these small variations in intensity after the edge.

EXELFS (the extended energy-loss fine structure) carries information about the bonding and co-ordination of the atoms contributing to the edge. However, the signal needs to be strong before statistically reliable information can be obtained.

Bruley *et al.* (1999) have employed this technique to study segregation-induced grain boundary embrittlement of copper. Changes in the ELNES permits correlation to be made between the presence of certain segregants and changes in the bonding of the atoms on the grain boundary. They demonstrated that Bi and Sb, well-known embrittlers of Cu, induce consistent changes in the bonding of Cu, while Ag, a similar segregant which does not embrittle Cu, produces no detectable changes in the bonding.

These authors produced TEM samples of Bi-doped, Sb-doped and Ag-doped copper foils, thinned to electron transparency using conventional preparation procedures. In all cases the presence of impurity segregation was confirmed using conventional X-ray energy-dispersive spectrometry. The EELS measurements were carried out with a STEM operating at 100 keV, with a nominal probe size of ~ 1 nm (full width at half maximum) with a current of about 0.5 nA. The conditions required to optimize detection sensitivity for interface analysis require the highest current density and are not consistent with achieving the smallest probes.

5.5.3.3. Bi-doped Cu. The most significant changes in the Cu $L_{2,3}$ edge were for a boundary exhibiting close to a monolayer of Bi, and is illustrated in Figure 5.42. The difference between that spectrum from the boundary and one from the matrix are made obvious in Figure 5.42(a) by overlapping the spectra. It is easy to discern the extra intensity in the edge-threshold portion of the spectrum for Bi-containing regions.

The difference between these two spectra is shown in Figure 5.42(b). The shape of the difference spectrum depends on the magnitude of the scaling factor used to normalise the two spectra. These authors chose a factor so as to match the integrated intensities in a 50 eV window centred at 980 eV, ~ 50 eV above the edge threshold.

5.5.3.4. Sb-doped Cu. An increased edge intensity in the presence of Sb is similar to that of Bi (Figure 5.43).

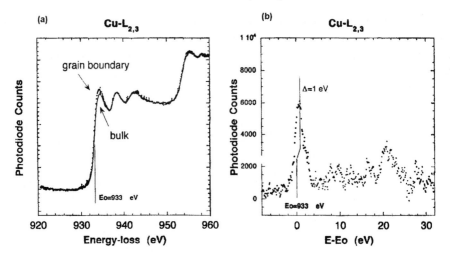

Figure 5.42. (a) shows the Cu $L_{2,3}$ edge of a Cu grain boundary containing close to a monolayer of Bi, compared with the spectrum recorded in the nearby matrix. A small but detectable increase in the edge intensity near the threshold at $E_0 = 933$ eV is apparent. (b) Shows the spatial-difference spectrum determined by subtracting the scaled matrix spectrum from the boundary spectrum. (After Bruley *et al.* 1999.)

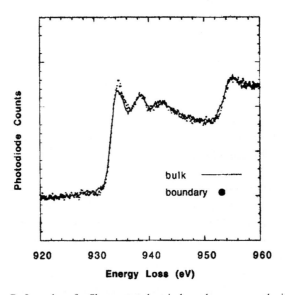

Figure 5.43. The Cu $L_{2,3}$ edge of a Sb-segregated grain boundary compared with the Cu matrix. (After Bruley *et al.* 1999.)

5.5.3.5. Ag-doped Cu. In all the boundaries examined by Bruley *et al.* (1999), Ag segregation did not lead to any observable effect on the Cu $L_{2,3}$ edge, either in the as-recorded spectra or the difference spectra. Ag segregation does not embrittle Cu, and so the absence of a detectable effect is consistent with the suggestion that electronic factors are responsible for grain boundary weakness.

5.5.4 Elemental Mapping Using EELS

In the same way that X-ray elemental maps can be formed from thin specimens using EDS techniques, EELS spectra can also be employed to map the distribution of selected elements present in the sample. The large background beneath most energy loss edges means that the contrast in such maps will be low, with a corresponding large detection limit. However, the availability of powerful computation facilities may permit the collection of a full spectrum from each point on the map. Each spectrum may later be processed to subtract the background before displaying the image. By reprocessing the data corresponding to another element, its distribution may in turn be mapped.

5.5.4.1. Energy Filtered Images. If a series of slits and lenses are introduced into the TEM column *after* the conventional EELS spectrometer, it is possible to reconstruct the image using only the electrons which have been selected by the spectrometer. The Gatan Imaging Filter (GIF) is one device which will perform this function. This approach has been used by Pénisson and Vystavel (2000) to study the penetration of nickel along two symmetrical [101] tilt boundaries in Mo bicrystals of different energy. The identification of the nanophases present at the grain boundary was performed using a combination of techniques including that of EELS spectroscopy, including the chemical imaging of the precipitates by the use of GIF.

Mo bicrystal slices of 1 mm thickness were cut by spark erosion, then mechanically thinned to a thickness of 0.1 mm. Discs of 3 mm diameter were then prepared and placed in the experimental assembly illustrated in Figure 5.44.

The 3 mm disc was covered by a 3 μm thick polycrystalline nickel foil, then a polycrystalline molybdenum disc was placed on top of the nickel foil. This composite sandwich was placed between two alumina supports and heated for 30 min at 1350°C. The cross-sections used for microanalysis were prepared by cutting with a wire saw in a direction perpendicular to the grain boundary, then mechanically and chemically polished.

In places where the orientation of the boundary changed, small precipitates were observed. Their EELS spectrum (Figure 5.45) showed that they had a composition corresponding to the NiMo-δ phase.

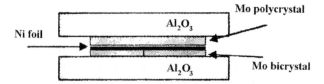

Figure 5.44. Experimental assembly used for the thermal treatment of Mo bicrystals in the presence of nickel. (Reproduced with permission of Pénisson and Vystavel 2000.)

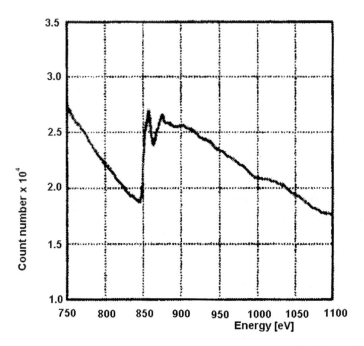

Figure 5.45. Electron energy loss spectrum of MoNi-δ precipitate showing the Ni L_{23} edge at 855 eV. (Reproduced with permission of Pénisson and Vystavel 2000.)

The chemical image obtained using the Ni L_{23} (855 eV) is shown in Figure 5.46. The shape of the precipitate is triangular, and one of the interfaces Mo–MoNi was found to be parallel to a set of {110} atomic planes in one of the two Mo crystals.

5.6. HIGH RESOLUTION ELECTRON LOSS SPECTROSCOPY (HREELS)

EELS can be applied in conjunction with several techniques. We have just discussed its use in association with TEM or STEM, where the losses predominantly occur in

Figure 5.46. Chemical image of the precipitate. (Reproduced with permission of Pénisson and Vystavel 2000.)

the bulk of the sample, as the electron beam passes through the thin specimen to the EELS detector on the other side.

Another technique is spin polarised EELS, or SPEELS, which can provide information on phenomena such as magnetic coupling and exchange excitation processes, but SPEELS will not be discussed further here.

In *surface analytical* techniques, a low energy electron beam of energy in the range 1 to 10 eV impinges on the sample surface. Small amounts of energy are lost, typically in inducing vibrations in molecules adsorbed on the surface. The beam is reflected from the surface resulting in a sharp peak corresponding to elastically scattered electrons, with a number of peaks at lower energy which correspond to plasmon or other excitations. In HREELS, the energy losses are examined at high resolution (about 30 meV), and data concerning the vibrations of molecules on surfaces can be determined, the technique provides complementary information to Reflection Absorption Infra Red Spectroscopy (RAIRS) and it can be considered as the electron-analogue of Raman spectroscopy. HREELS is sometimes referred to as VELS (vibrational ELS). RAIRS has the advantage of significantly greater energy resolution, but the energy range that can be studied is limited by the availability of suitable sources, whereas HREELS can study vibrational features of energy right down to a few meV. These aspects of the technique are reviewed by Ibach and Mills (1982). A new variant, TREELS (Time-resolved Electron Energy Loss Spectrsocopy) allows one to monitor these signals in real time to study kinetic events.

5.6.1   Instrumentation

Since the technique employs low energy electrons, it is necessary to use a UHV environment. The high energy resolution in the incident electron beam is achieved by monochromatizing a thermionic electron source by means of a CHA. A second CHA is used as an energy analyser, and the basic experimental geometry is as illustrated schematically in Figure 5.47.

The highly monoenergetic beam of electrons is directed towards the surface, and the energy spectrum and angular distribution of electrons backscattered from the surface are measured. The kinetic energy of the incident electrons is typically in the range of a few electron volts, and under these conditions the electrons penetrate only the outermost three or four atomic layers of the crystal. High resolution is achieved through the production of a very mono-energetic and well-collimated electron beam, together with the use of sensitive detectors. Without degrading the signal below the level of detectability, the best energy resolution that is presently obtainable is 3.7 meV. In order to detect weak signals, however, it may be necessary to operate with lower resolution.

5.6.2   Spectra

The interpretation of HREELS is similar to that of RAIRS (pp. 41 *et seq.*), although RAIRS spectra are usually expressed in wavenumbers, while HREELS are in meV ($1\,meV = 8.065\,cm^{-1}$). The resolution in RAIRS is typically 0.25 meV, while it is about 20 meV for HREELS (despite the HREELS acronym!).

A substantial number of electrons are elastically scattered, and this gives rise to a strong *elastic peak* in the spectrum. When an electron of low energy (2–5 eV) approaches a surface, it can be scattered inelastically by two basic mechanisms, and the data obtained are dependent upon the experimental geometry – specifically the angles of the incident and the (analysed) scattered beams with respect to the surface (θ_1 and θ_2 in Figure 5.47). Within a certain distance of the surface the incident electron can interact with the dipole field associated a particular surface vibration, e.g. either the vibrations of the surface atoms of the substrate itself, or one or other

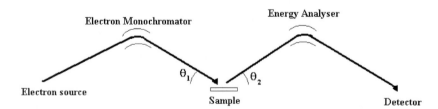

Figure 5.47.   Schematic illustration of an HREELS apparatus.

of the characteristic vibrations of an adsorbed molecule. *Specular geometry* is when $\theta_1 = \theta_2$, and scattering is principally by this long-range *dipole scattering*. *Off-specular geometry* is when $\theta_1 \neq \theta_2$, and electrons lose energy to the surface species by being scattered inelastically from an atomic core at the surface, again either of a substrate or of an adsorbate atom. This short-range mechanism is called *impact scattering*.

5.6.2.1. Dipole Scattering. Most HREELS observations have been concerned with this mechanism, where the oscillating dipoles on the specimen surface most often arise from the vibrational modes of the molecular absorbates that are present. The incoming charged electron is affected by the vibrating dipole at the surface, and the energy loss of the specularly scattered electron is characteristic of the energy deposited in the vibrational mode (exactly as in the case of an IR spectrum). The scattering process is regarded as taking place by either of two paths – an inelastic reflection followed by interactions with the vibrating dipoles through the long-range Coulomb field, or vice versa.

There is also the 'normal dipole selection rule' in operation, as illustrated in Figure 5.48, due to Lüth (1981). Any dipole at a surface induces an image charge within the surface. If the dipole orientation is normal to the surface, the effect is enhanced by the image dipole. If, however, the orientation is parallel to the surface, the effect is annihilated by the image dipole. This orientation selection rule thus strongly favours normally oriented dipoles.

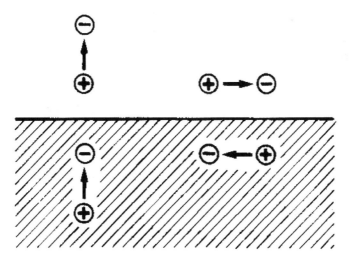

Figure 5.48. Schematic illustration of the operation of the normal dipole selection rule in HREELS. (After Lüth 1981.)

5.6.2.2. Impact Scattering.   This short range scattering process is from the ion core, and although the scattering is more isotropic (i.e. not only in the specular direction) the energy losses still reflect vibrational excitations in the absorbate. The angular distribution of peaks around the specular direction can distinguish between peaks which result from differing scattering modes, although dipole scattering is dominant.

As indicated in Figure 5.49, a substantial number of electrons are elastically scattered, giving rise to the strong *elastic peak* in the spectrum. On the low kinetic energy side of this main peak, (i.e. in the direction of increasing energy loss) additional weak peaks are superimposed on a mildly sloping background. One such peak appears in Figure 5.49, and such peaks correspond to electrons which have undergone discrete energy losses during the scattering from the surface. In Figure 5.49 the data are plotted against the energy loss, and the magnitude of this loss is equal to the energy of the vibrational mode of the adsorbate excited in the inelastic scattering process. A measure of the instrumental resolution is given by looking at the FWHM (full-width at half maximum) of the elastic peak.

In spectral identification, the first step is a comparison of the observed losses with vibrational frequencies measured by IRS in the gas phase, to see if any correlations exist. When a molecule is attached to a surface it is fettered by forces due to the chemical bonds to the surface, and there will be 'stretching' modes of vibration

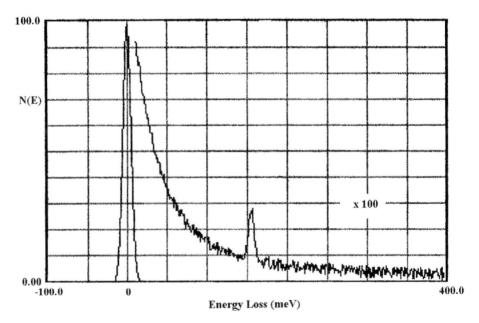

Figure 5.49.   A typical HREELS spectrum.

additional to those seen in the free molecule. Where a complex polyatomic molecule is adsorbed on a surface at several adsorption sites, the resultant spectrum can be very complex. These problems are discussed by, e.g., Rivière (1990).

5.6.3 Applications of HREELS

One of the classic examples of an area in which vibrational spectroscopy has contributed to the understanding of the surface chemistry of an adsorbate is that of the molecular adsorption of CO on metallic surfaces. Adsorbed CO usually gives rise to strong absorptions in both the IR and HREELS spectra at the (C–O) stretching frequency. The metal–carbon stretching mode ($\sim 400\,\mathrm{cm}^{-1}$) is usually also accessible to HREELS.

The vibrational modes of molecules adsorbed on a surface provide one with direct information on the nature of the chemical bonds between a molecule and its substrate.

REFERENCES

Auger, P. (1925) *J. Phys. Radium*, **6**, 205.
Briggs, D. & Seah, M.P. Eds. (1990) *Practical Surface Analysis* (Second Edition) Vol. 1, John Wiley & Sons, Chichester.
Bruley, J., Keast, V.J. & Williams, D.B. (1999) *Acta Materiala*, **47**, 4009–4017.
Carpenter, D.T., Wanatabe, M., Barmak, K. & Williams, D.B. (1999) *Microsc. Microanal.*, **5**, 254–266.
Cliff, G., & Lorimer, G.W. (1975) *J. Microscopy*, **103**, 203.
Egerton, R.F. (1986) *Electron Energy-Loss Spectroscopy in the Electron Microscope*, Plenum Press, New York.
Flewitt, P.E.J. & Wild, R.K. (1985) *Microstructural Characterisation of Metals and Alloys*, The Institute of Metals, London.
Goldstein, J.I. & Yakowitz, H. (1975) *Practical Scanning Electron Microscopy*, Plenum Press, New York.
Goldstein J.I., Newbury, D.E., Achlin, P., Joy, D.C., Romig, A.D.Jr., Lyman, C.E., Fiori, C.E. and Lifshin, E. (1992) *Scanning Electron Microscopy and X-ray Microanalysis.* (second edition), p. 341, Plenum Press, New York.
Heatherly, L. & George, E.P. (2001) *Acta Materiala.*, **49**, 289–298.
Hörz, G. & Kallfass, M. (2000) *Materials Characterization*, **45**, 391.
Ibach, H. & Mills, D.L. (1982) *Electron Energy Loss Spectroscopy and Surface Vibrations*, Academic Press, New York.
Joy, D.C. (1981) Electron energy loss, in '*Quantitative Microanalysis with High Spatial Resolution*', Metals Society, London, 277.
Li, D.X., Rao, J.C., Zhou, Z.X., Peng, H.Y., & Lan, J.Z. (2000) *Materials Characterization*, **44**, 391–402
Lüth, H. (1981) *Festkörperprobleme*, **XXI**, 117.

Lyman, C.E. (1987) *Physical Aspects of Microscopic Characterization of Materials*, Kirschner, J., Murata, K., & Venables, J.A. Eds., *Scanning Microscopy International* [suppl. 1], pp.123–134. (AMF, O'Hare, IL.)

Matsuya, M., Fukada, H., Kawabe, K., Sekiguchi, H., Inagawa, H. & Saito, M. (1988) *Microbeam Analysis*, 329.

Michael, J.R. (1981) *Practical Analytical Electron Microscopy in Materials Science*, second Edition, Ed. Williams, D.B. p.83, Philips Electron Optics Publishing Group, Mahwah, NJ, 1987.

Newbury, D.E., Wollman, D.A. Irwin, K.D., Hilton, G.C. & Martinis, J.M. (1999) *Ultramicroscopy*, **78**, 73–88.

Newbury, D.E. & Williams, D.B. (2000). *Acta mater.*, **48**, 323–346.

Overwijk, M.H.F., van den Heuvel, F.C. & Bulle-Lieuwma, C.W.T. (1993) *J. Vac. Sci. Technol.*, **B11**, 2021–2024.

Pénisson, J.M. & Vystavel, T. (2000). *Acta Materialia*, **48**, 3303–3310.

Prutton, M., Larson, L.A. & Poppa, H. (1983) *J. Appl. Phys.*, **54**, 374

Reed S.J.B. (1966) ICXOM 6, 339.

Reed, S.J.B. (1993) *Electron Microprobe Analysis* (Second Edition) Cambridge University Press, Cambridge UK.

Rivière, J.C. (1990) *Surface Analytical Techniques*, Clarendon Press, Oxford.

Romig Jr., A.D. & Goldstein, J.I. (1979) *Microbeam Analysis*, Newbury, D.E. Ed. pp. 124–156, San Francisco Press, New York.

Seah, M.P. (1980) *J. Vac. Sci. Technol.*, **17**, 16.

Springer, G. (1976) *X-ray Spectrom*, **5**, 88.

Suzuki, S., Abiko, K. & Kimura, H. (1981) *Scripta Metallurgica*, **15**, 1139–1143.

Watanabe. M., Horita, Z. & Nemoto, M. (1996) *Ultramicroscopy*, **65**, 187–198.

Watanabe, M. & Williams, D.B. (1999) *Ultramicroscopy*, **78**, 89–101.

Williams, D.B. & Carter, C.B. (1996) *Transmission Electron Microscopy: a Text for Materials Science*, Plenum, New York.

Williams, D.B., Watanabe, M. & Carpenter, D.T. (1998) *Mikrochim Acta* **15**, (Suppl.) 49–57.

Williams, D.B., Papworth, A.J. & Watanabe, M. (2002) *Journal of Electron Microscopy*, **51** (Supplement), S113–S126.

Wollman, D.A., Nam, S.W., Hilton, G.C., Irwin,K.D., Dulcie, L.L., Rudman, D.A., Martinis, J.M. & Newbury, D.E. (2000) *J. Micros.*, **199**, 37–44.

Zieba, P., Cliff, G., & Lorimer, G.W. (1997) *Acta Mater.*, **45**, 2093–2099.

Ziebold, T.O. (1967) *Anal. Chem.*, **39**, 858.

FURTHER READING

Goodhew, P.J., Humphreys, F.J. & Beanland R. (2001) *Electron Microscopy and Analysis* (Third Edition), Taylor and Francis, London and New York.

Kohler-Redlich, P. & Mayer, J. (2002) Quantitative Analytical Transmission Electron Microscopy, in '*High-Resolution Imaging and Spectrometry of Materials*' Ernst. F., & Rühle, M., Eds., pp. 119–188, Springer, Berlin.

Chapter 6
The Choice of Technique

Chapter 6
The Choice of Technique

We will now take the opportunity of reviewing briefly the range of techniques of local analysis that we have discussed in the preceding chapters, drawing attention to the particular strengths and weaknesses of each. This should assist a practising materials scientist to select a method of analysis most appropriate for the solution of a problem he or she may face.

The use of vibrational spectroscopy for the qualitative analysis of absorbed surface species is first considered, and a Table is then included which summarises a number of the key features of the various quantitative techniques. We then proceed to summarize these in groups depending not upon the probe used (as in the preceding chapters), but in terms of the signal emitted by the specimen which is used in each identification process.

We have not included Atom Probe Microanalysis in this scheme. It constitutes the ultimate in local analysis – in that individual atoms can be selected and identified by TOF spectroscopy. Chapter 1 gives an account of the range of applications of the technique at the present time: the development in atom-probe methods has allowed the continuing increase of both the volume of material that can be mapped at the atomic scale and the quality of the data obtained.

6.1. SURFACE ANALYSIS

6.1.1 Vibrational Spectroscopy
This is a standard method of *identifying molecular species* adsorbed on a surface, as well as species generated by surface reaction. We have considered three such techniques.

IR spectroscopy is a widely used non-destructive tool for in *in situ* studies of catalysis and at electrode surfaces of electrochemical cells. The spatial resolution is comparable to the wavelength of the light, i.e. < 1µm to tens of micrometres. The exact frequency of the absorbed light provides a characteristic signature of the molecules, ions, or radicals present in the sample, and good fingerprint libraries are available to provide reliable qualitative analysis. Sampling may be difficult in some cases, and mid-IR light does not penetrate many common optical materials. In addition, water absorbs mid-IR light strongly, so aqueous samples may be probed only as thin films.

Raman microscopy has the ability to investigate regions down to 1 µm, and, by the aid of fibre optics, remote sampling is possible. The molecular information obtained may complement that provided by other microanalytical techniques. The probe is water-compatible, sampling is easy and is often non-invasive. The Raman microprobe may be used to investigate organic substances as well as inorganic compounds; only very small samples are required, and small local heterogeneities can be detected in larger samples. Depth profiles may be obtained directly in transparent materials, otherwise microtomed sections can be studied by a lateral scan.

HREELS employs a source of low-energy electrons, and it provides information complementary to RAIRS. It can be considered as the electron-analogue of Raman spectroscopy. Although the technique has a low energy resolution ($\sim 20\,$meV compared with $0.25\,$meV in RAIRS), HREELS can study vibrational features of energy to lower energy values than in RAIRS, i.e. to below $400\,cm^{-1}$. It has, for example, contributed to the understanding of the surface chemistry of the molecular adsorption of CO on metallic surfaces where the metal–carbon stretching mode is $\sim 400\,cm^{-1}$ (N.B. $1\,meV = 8.065\,cm^{-1}$).

6.1.2 *Electron, Ion, and Electromagnetic Radiation Spectroscopies*

These techniques are generally able to provide quantitative as well as qualitative analysis of the specimen surface, and a number of them may be used in combination to maximize the amount of information obtained.

Their characteristics are summarized in Table 6.1.

We will consider the techniques in turn.

X-ray photoelectron spectroscopy is frequently applied in the fields of catalysis and polymer technology. It has poor spatial resolution, and is generally limited to homogenous samples. Radiation sensitive materials are more appropriate for XPS analysis, as the X-ray beam is less damaging to the specimen surface than the electron beam used in AES, partly due to the lower flux densities that are used.

Depth profiling may be conducted on insulating materials, and XPS can perform thickness measurements on organic films such a lubricants on magnetic media. It is capable of measuring surface contamination by both inorganic and organic materials on a wide variety of substrates.

Auger electron spectroscopy is preferred over XPS where high spatial resolution is required, although the samples need to be conducting and tolerant to damage from the electron beam. Many oxides readily decompose under electron radiation, and this may give rise to difficulty in spectral interpretation, and this has restricted the application of AES in the field of catalysis.

In its basic form, a broad-focussed electron beam probe provides compositional information on a relatively large area ($\sim 1\,mm^2$), to a depth of the order of 1 nm. To obtain information about the variation of composition with depth, controlled surface etching of the analysed region is achieved by simultaneously exposing the surface to an ion flux which leads to sputtering of the surface atoms.

In SAM the electron beam can be focussed to provide a spatial resolution of < 12 nm, and areas as small as a few micrometers square can be scanned, providing compositional information on heterogeneous samples. For example, the energy resolution is sufficient to distinguish the spectrum of elemental silicon from that of silicon in the form of its oxide, so that a contaminated area on a semiconductor device could be identified by overlaying the Auger maps of the two forms of silicon obtained from such a specimen.

EELS analysis has great sensitivity to the lighter elements ($Z < 10$), and it has the great attraction that it can be used to detect He, Li and Be, for example, which are otherwise very difficult to analyse. It is capable even of detecting hydrogen. Qualitative analysis is possible by comparing the energies of any edges in the energy-loss spectrum with tabulated values for the elements.

The energy loss near-edge structure of the spectrum permits the chemical form of a given element to be identified. Thus, on pp. 191 *et seq.* we have given an example of using the changes in the ELNES to correlate the presence of certain grain boundary segregants with the bonding of the atoms on the grain boundary.

EELS spectra can also be employed to map the distribution of selected elements present in a sample, in the same way that X-ray elemental can be exhibited in analytical TEM experiments.

6.1.2.1. Analysis by the Detection of Scattered Ions. Ions generally penetrate the specimen much less deeply than electrons of equivalent energy, so they are more surface-sensitive. Ion-based surface analytical techniques are popular because of their sensitivity and their ability, in some cases, to reveal the depth composition profile.

The *laser microprobe* volatilizes a small region of the sample, and released ions are analysed in a TOF mass spectrometer. It provides a rapid elemental survey analysis, and among its applications is the mass analysis of single particles – notably those encountered in the airborne state. As in the case of SIMS, quantitation is hampered by the fact that the ionization probability is matrix dependent, although different isotopes can be distinguished. When the technique was first developed, its high sensitivity, mass range and isotope discrimination overcame several limitations of other well-established microprobing techniques such EPMA.

SIMS has the principal advantage of high surface sensitivity (only a very small proportion of the detected ions come from the second or lower layers of the material being analysed) and its very low detection limits for impurities. All elements can be detected at concentrations of ppm ($\sim 0.0001\%$) in the surface region. This may be contrasted with AES or XPS where the detection limit is $\sim 0.1\%$ with a sampling

Table 6.1. A comparison of local analysis techniques

Technique	Detection	Lateral resolution	Min. conc. detected	Element range	Chemical state info.	Quantitative
XPS	Electrons	Imaging XPS: < 10 μm	0.01–1 at%	All except H and He	Yes	Yes (Standards)
AES	Electrons	< 12 nm by SAM	0.1–1 at%	All except H and He	Yes	Yes using standards
EELS	Electrons	Atomic level	Atomic level	Esp. $Z < 10$	Yes	Yes
Laser microprobe	Ions	0.6–1.5 μm	10^{-19} g	All elements	Yes	No
SIMS	Ions	20 nm–1 μm depending on ion source	Ppm for most elements Ppb for most favourable	All elements; isotopes distinguishable	Yes	Not in general, but internal standards usable in some cases.
RBS	Ions	10^{-1} mm^2 area	1–10 at% ($Z < 20$) 0.01–1 at% (20 < Z < 70) 0.001–0.01 at% ($Z > 70$)	All except H and He	No	Yes (No need for standards)
HIBS	Ions	As for RBS	$< 10^{10}$ atoms/cm^2	All elements above Ar in a single spectrum	No	Usually not
ERDA (HFS)	Ions	~0.1–1 mm depending on beam size	0.01 at%	Light samples ($Z < 9$) e.g. Hydrogen isotopes	No	Yes
EPMA	X-rays	~2 μm	0.1–1 at%	Be to U	No	Yes
FEGSTEM	X-rays	<1–2 nm	~10^{-22} g	$Z > 4$	No	Yes
PIXE	X-rays	0.1–1 μm	~0.001% with micro-PIXE	Al (Z=13) to U	No	Yes (No need for standards)
NRA	X-rays, gamma rays, ions	A few microns	A few ppm	For $Z < 9$; isotope specific	No	Yes (No need for standards)
CPPAA	Gamma rays	As for NRA	10^{-2} μg/cm^2		No	Yes
PIGE	Gamma rays	~1 μm	Li: 10–100 ppm F: 1–10 ppm B: 500–1000 ppm	Light elements ($Z < 15$)	No	Yes

Table 6.1. continued

Technique	Depth	Depth resolution	Depth profile	Insulators	Vacuum	Composition maps	Beam diameter
XPS	1.5–4.0 nm	1–10 nm	Yes by sputtering	Yes with flood gun	UHV	Yes by imaging XPS	< 10 μm possible and up to 2 mm
AES	~1 nm	< 1 nm	Yes by sputtering	No	UHV	Yes, by SAM	10 nm possible
EELS	< 1 μm	N/A	No	Yes	Yes	Yes	As for FEGSTEM
Laser microprobe	~1 μm	N/A	No	Yes	Yes	Yes	~0.5 μm
SIMS	Up to 10 μm by dynamic SIMS	2–5 nm possible; 10–20 nm typical	Yes by dynamic SIMS	Yes	Yes	Yes, by imaging SIMS	1 μm imaging 30 μm depth profiling
RBS	~1–2 μm to 20 μm	20–30 nm; 3–4 nm with tilted targets	Yes up to a few microns below surface.	Yes	UHV	Yes	~2 μm with microbeam, otherwise 2 mm
HIBS	~10 nm	N/A	No	Yes	Yes	No	~1 mm
ERDA (HFS)	~1 μm	10 nm close to the surface; 30–80 nm elsewhere	Yes	Yes	Yes	No	~1 mm
EPMA	~1 μm	1–5 μm	No	Yes, by coating	Yes	Yes	0.2–1 μm
FEGSTEM	~1 μm	N/A	In cross-sections	No	Yes	Yes	< 1 nm
PIXE	Several tends of microns	N/A	In cross-sections	Yes	Yes	Yes	1–10 μm in micro-PIXE
NRA	A few microns	A few nm to tens of nm	Yes, especially for ultra-shallow depths	Yes	No	No	~4 mm
CPPAA	Typically 0.1 to 1 mm	~2%	Not possible	Yes	No	No	As for NRA
PIGE	~100 μm		In cross-sections	Yes	Yes	Yes	~5 mm

depth of ~ 5 atom layers. In general this difference arises because the peak-to-background ratio for secondary effects excited by ions is larger than that for electron excitations. The spatial resolution of SIMS is limited by the diameter of the incident ion beam, and this is usually worse than that of electron probes.

Static SIMS is appropriate for obtaining information on the lateral distribution of surface chemical species. A broad, defocussed ion beam is often used in order to minimise surface damage. In dynamic SIMS sample erosion takes place quite rapidly, and depth profiles are obtained by monitoring peak intensities in the mass spectrum of sputtered ions as bombardment proceeds.

RBS is a quantitative analytical tool which provides simultaneously the depth profile and the composition by mass number of the sample. The disadvantage is that a large and expensive particle accelerator is required to produce the incident beam. The probe depth of RBS is typically 1–2 μm with a depth resolution of 20–30 nm.

By adding accessories to the sample chamber, or by changing the operating procedures, several other experiments can 'piggy-back' on to the RBS analysis. For example HIBS, HFS, PIXE, NRA, CPAA and PIGE may all be accessible using a given particle accelerator.

It is possible to measure nearly any type of sample for almost any element with little or no preparation. Only a few mg of sample is required, and the measurements are non-destructive in that the sample is generally undamaged. Measurements take only 1–20 min of beam time. Elemental mapping showing the variations in elemental concentrations can be measured over the surface of a sample using the ion microprobe for an area as large as 5×5 mm.

HIBS uses ions heavier than He^{++}, and collision cross-sections are higher for heavy primary ions. These heavy ion beams provide advantages in trace heavy element determinations of light element samples, and they also have superior mass resolution for elements in the same row of the Periodic Table in the backscattered spectroscopy enabling (for example) the separation of Ga and As signals from thin films of GaAs. The matrix elements are all scattered forward and cannot contribute interference signals. The accelerators and detectors are the same for HIBS and RBS, but the back-scattering cross-section and kinematic factor calculations for HIBS are of uncertain validity, so the data are generally limited to qualitative analysis.

The HIBS system was invented and patented in the USA at Sandia National Laboratories, and it has been transferred to the University of Central Florida in Orlando, USA. It was brought into operation there in December 2001.

ERDA (HFS) only requires the addition of a thin foil (of carbon, mylar or aluminium) to separate forward scattered hydrogen from forward scattered primary He^{++} ions. The analytical information obtained consists of hydrogen concentration versus depth. The sample is tilted so that the He^{++} beam strikes at a grazing angle, giving a HFS depth profile resolution of about 50 nm. The surface hydrogen content is determinable within an accuracy of 5% with a detection limit of 0.01%. An example of a heavy-ion ERDA experiment is given in Chapter 4.

6.1.2.2. Analysis by the Detection of X-rays or γ rays. *EPMA* is a fully qualitative and quantitative method of non-destructive analysis of micrometre-sized volumes at the surface of materials, with sensitivity at the level of ppm. All elements from Be to U can be analysed, either in the form of point analysis, from line scans and also as X-ray distribution maps. Current software allows the combination of elemental data in the latter, so that, for example, the digital data for those elements that corresponds to a selected phase will produce an X-ray map of the distribution of that phase in a given microstructure.

PIXE employs excitation of the sample by means of charged α particles from an accelerator, and the emitted X-rays are analysed in a multichannel analyser which gives a pulse height spectrum of the X-rays. These heavy particles give rise to less background in the recorded spectra than, for example, electron or X-ray excitation, hence the characteristic X-ray peaks stand out more clearly, giving better sensitivity. The sample composition can be determined in absolute terms, and although PIXE analysis can in principle cover the whole periodic table, it is difficult to determine the lightest elements because their characteristic X-rays have such a low energy that they are easily absorbed. This means that the sensitivity drops rapidly with decreasing atomic number, and the practical lower limit is around aluminium ($Z = 13$) so that lighter elements are usually not determined by means of PIXE.

For the analysis of large objects which cannot be placed within the irradiation chamber it is possible take the particle beam into the ambient air through a thin window at the end of the beam line. In this way any type of object can be analysed – for example paintings and archaeological artefacts.

The PIXE microbeam technique has a spot size in the range 1–10 µm, and this enables a study of the spatial distribution of elemental concentrations. The advantage of µ-PIXE over EPMA is a very much increased analytical sensitivity due to the much lower Bremsstrahlung background generated by the proton beam. The detection limits are of the order 0.1% for EPMA and 0.001% using the µ-PIXE technique.

NRA is a powerful method of obtaining concentration versus depth profiles of labelled polymer chains in films up to several microns thick with a spatial resolution of down to a few nanometres. This involves the detection of gamma rays produced by irradiation by energetic ions to induce a resonant nuclear reaction at various depths in the sample. In order to avoid permanent radioactivity in the specimen, the energy of the projectile is maintained at a relatively low value. Due to the large coulomb barrier around heavy nuclei, only light nuclei may be easily identified (atomic mass ≤ 30).

In Chapter 4.7 we have given examples of the application of the technique to both polymers and to silicon implanted with P. In the latter case it appears that NRA has a better reproducibility than SIMS in the determination of ultrashallow dopant profiles. Demortier (2000) also presents examples of applications of NRA to archaeological and biological materials.

CPAA measures the characteristic decay radiation of the radionuclides produced by the incident charged particles. The technique has been widely applied in the determination of trace elements concentrations in bulk samples, but it also has possibilities for surface characterisation, provided the thickness of the layer to be characterised is less than the range of the charged particles employed.

The main advantages of CPAA as a surface characterisation method are that it requires no sample preparation, its high accuracy and its low detection limits. It is an independent method, because no standard samples calibrated by other methods are needed, so that its value may lie in its ability to calibrate standard samples for other, more routine analytical methods.

PIGE is often available in a PIXE setup by placing a germanium detector outside the irradiation chamber, but as close as possible to the sample. Since the same particle beam may be used in both cases, PIXE and PIGE analyses can be performed simultaneously. PIGE is a rapid, nondestructive technique that could, in principle, be used to analyse any element, but in practice is employed in the analysis of light elements (e.g. Li, B and F) which are often difficult to determine by other analytical techniques. The sensitivity of the technique varies from isotope to isotope, and the sample matrix will also influence the detection limit of an element because of variations in the background of the gamma-ray spectrum.

External microbeam scans may be conducted in air, if analyses *in vacuo* are difficult to perform on a given specimen. The thin specimen is placed $\sim 100\,\mu m$ from the exit foil of the microbeam, and the gamma-ray detector is placed just behind the sample.

6.2. THE ANALYTICAL TEM

The development of X-ray microanalysis in the TEM has been driven by the improvement in spatial resolution in comparison with EPMA. This arises because thin specimens are used, so less electron scatter occurs as the beam traverses the specimen, and also because of the higher electron energy in the TEM also reduces scatter. The disadvantage is that the specimen has to be prepared in the form of a thin foil, and the problems involved in this process have already been discussed.

The electron probe can be from $< 1\,nm$ to $10\,nm$ in diameter, and positioned to localize the analytical signals. To obtain the smallest electron probe, a field-emission gun (FEG) is essential, and microanalysis is accomplished using an XEDS and/or an

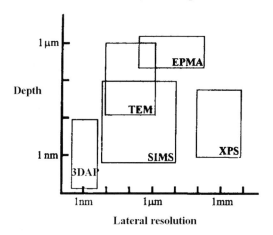

Figure. 6.1. A map showing the origin of the analytical information for a number of the techniques discussed.

EELS interfaced to the TEM column. Elements with an atomic number $Z < 4$ cannot be detected, and quantification of the signal from elements with $Z < 8$ is difficult. The accuracy of the quantification procedure in the TEM is about $\pm 5\%$, compared with EPMA (about ± 1–2%), although the actual mass analysed can be very small ($\sim 10^{-22}$ g).

The strength of analytical TEM is that compositional information, images, and diffraction data can be simultaneously acquired from the same area of the specimen with a spatial resolution of < 1–2 nm.

6.2.1 Comparison of Techniques

Figure 6.1 is a graphical comparison of several analytical techniques, plotting the depth from the surface from which the analytical information comes versus the lateral resolution. A fuller summary of the characteristics appears in Table 6.1.

It is obvious that no one technique can solve all problems: many approaches are needed and I hope that the preceding chapters will guide the interested reader to adopt the most appropriate technique for each problem he or she faces.

REFERENCES

Demortier, G. (2000) Nuclear Reaction Analysis in '*Encyclopedia of Analytical Chemistry*', John Wiley & Sons, Chichester.

Subject Index

Materials Index